The Infantry Regiments of Frederick the Great 1756-1763

Günter Dorn   Joachim Engelmann

Schiffer Publishing Ltd

1469 Morstein Road, West Chester, Pennsylvania 19380

Translator's Note

In the translation of this book and its index, all German personal names have been left in German with one exception: "Friedrich der Grosse" has been translated as "Frederick the Great", but Friedrich has remained Friedrich when otherwise referred to. Similarly, German place names have been left in German unless a well-known English form exists, such as "Prussia" for "Preussen". Titles of nobility have been translated, though this removes the distinction between "Fürst", a ruling prince, and "Prinz", a non-ruling member of a ruling family. Non-German names have generally been rendered in the form best known to English-speaking readers. It must be pointed out that the authors seem to have converted the names of numerous non-German persons, particularly French noblemen, into German, and these have been left in German.

Translated from German by Dr. Edward Force.

Library of Congress Catalog Number: 88-64005.

Printed in the United States of America.
ISBN: 0-88740-163-5
Published by Schiffer Publishing, Ltd.
1469 Morstein Road, West Chester, Pennsylvania 19380

This book may be purchased from the publisher.
Please include $2.00 postage.
Try your bookstore first.

CONTENTS

REGISTER OF NAMES

PREFACE

When the early evening of December 5, 1757 kindheartedly covered in darkness the cemetery of Leuthen, covered with dead and wounded men, there sounded, first tremulously, echoing the day's grim reality, then firmly, with manly strength, from the mouth of an unknown young grenadier, the chorale "Now Let Us All Thank God". The melody was taken up by his comrades, and soon the hymn of praise resounded movingly from the mouths of the whole army.

What kind of an army was it that thanked God in this way for a victory gained "against all the rules of warfare", instead of—as happened more often than suited the victims, even in the thoroughly human era of the Enlightenment—robbing and plundering, as victors always did? The people of our time are very quick to give flatly negative judgments, even when they understand as much of history as the inhabitants of a great metropolis know of Alpine agriculture. They talk of inhuman officers, of unbearable social conditions, of extreme material deprivation, of mistreatment and oppression, but the proponents of this opinion are not in a position to universalize the thoroughly real negative individual incident. All the same, Wolf Schneider stated in his "Book of the Soldier": "In the Russian Army and the English Navy it was worse, to be sure, but Prussia has certainly had the worst press", but who pays any attention to such casual remarks? Even those whom one can regard as cultivated in the customary sense, and who can cite from contemporary literature, keep on quoting the same meager sources that come nowhere near satisfying those who are really interested. The Swiss Ulrich Bräker, with his tear-jerking life story "The Poor Man from Tockenburg", is most Assuredly one of them. Although long forgotten, a fully insignificant professor from the first half of our century dug him out of the literary junk-room again in order to let the unprepossessing, simple-minded fellow from the isolated country of the Swiss Alps, whom Prussian recruiters had pressed into military service by devious means, take his delayed revenge on Prussia and its army. Since then this book by "the first plebeian author", who is characterized in Kindler's Literature-Lexicon by "scribbling and imagining", has served countless writers and directors of our times as a starting-point and basis for slandering the old Prussian Army and its king. Sometimes it is even amusing to see with what rigidity the media of our day and their masters avoid the countless positive but no less realistic portrayals of contemporaries. Nobody wants to know the Diary of the Junker von Barsewisch, Dreyer's "Life and Deeds of a Prussian Regimental Drummer", or the old Prussian military poetry of an Ewald von Kleist, Hölty or Gleim, and pick a few at random out of the unending richness of contemporary descriptions. Anyone to whom all this evidence from loyal Prussians is too extravagant may chart his own course to the truth between Bräker and Dreyer—down-to-earth contemporaries who experienced it all firsthand. Books like Augstein's "Prussia's Frederick and the Germans", written from the same vantage point of two hundred years later, are undoubtedly less capable of presenting a historically accurate picture. It was still opportune a few years ago to trample on the graves of the old Prussians.

But this does not yet answer the question of the nature of Frederick's army. In fact, this can scarcely be done in a few simple sentences. The well-remembered military scholar Jürgen Olmes made the best attempt when he wrote in the foreword to the diary of the Junker Ernst Friedrich Rudolf von Barsewisch, newly edited by him: "Never would the King, never would his commanders have won unwilting laurels if their regiments had not been sacrificing, loyal and devoted at the core. . . !" Let us particularly not forget the tremendous spiritual burdens to which the soldiers, above all the line infantry, were exposed every day. Keeping in close-ordered bodies or advancing under murderous small-arms and grapeshot fire, not being allowed to seek any shelter or leave their commanded place, with the gloomy certainty that even minor wounds, given the primitive state of war surgery at that time, could scarcely be survived at all, and then only as cripples, demanded great strength of spirit, trust in God and devotion to a cause that all too often was not that of the soldier himself.

Since this obvious quality of the Prussian Army cannot be denied to this day, one can always consider the oft-quoted cynical saying of Frederick the Great—in what forgettable volume is it found?—namely that the soldier must fear his officer more than the enemy. To be sure, there was strict discipline in the Prussian Army, but do beatings really help overcome the fear of death? Many officers, of course, never beat their men, and in a goodly number of regiments the commanders simply forbade the punishment of soldiers with the whip. The viewpoint of many officers can be found in the words of the Junker of Barsewisch: "I call it good fortune, but at the same time a lot depends on an officer's winning the esteem of his underlings and taking the best possible care of them, as the common soldier has such a difficult service, in that he is often deprived of all comforts and amenities, and can turn to no one but his officer with his little needs and requests." Barsewisch was not alone in having these human thoughts. The King himself, in his regulations for the infantry of 1743, which were in force for the total extent of his reign, urged his officers several times to treat their underlings justly and humanely. A junior officer who beat a man till he was unable to serve had to expect severe punishment. The King showed concern even for the soldiers who had to retire from service for reasons of age, contradicting the widespread false notion that he possessed no feeling for social responsibility. The regulations speak of this in eloquent terms: "When in the regiment fellows have become invalided because of age or illness, then His Royal Majesty absolutely does not wish them to be retired without it being made known in advance, rather the regiments shall inform His Majesty, so that such may be cared for and not need to go begging.

NB: Let it be recorded every year, so that the departing soldiers may keep a coat, shirt and trousers, so that the departing soldier need not leave the regiment naked."

The King's relationship to the individual soldier and the regiment depended on their achievements. In many cases the King had no understanding of failure. A regiment that had lost the good graces of its ruler could win them back only through extraordinary bravery before the foe and faultless performance when inspected. In fact, even that was often not enough. For as long as he lived, he never forgave the East Prussian regiments that had panicked at Zorndorf, and he "infamously cashiered" General von Finck, who had struck his colors in a hopeless situation at Maxen.

"It is noteworthy that deserters who joined foreign armies often spoke critically of the disgraceful conditions that prevailed there, and longed for the good old discipline. When in a free city of the German Empire one of Frederick's former junior officers made a small mistake in a drill, the colonel of the city militia wanted to criticize him before his men. The junior officer stepped forward and announced in a loud voice: 'Colonel, do not forget that I have served the King of Prussia!' Then he made a stiff about-face and marched back to his place. The colonel did not dare to reply." (Christopher Duffy, Frederick the Great and his Army)

Such was the spirit of an army whose nucleus and backbone, the line infantry, will be portrayed in words and pictures in this volume. The author of the book, Joachim Engelmann, is a renowned military historian and scholar who has become known inside and outside this country through a number of publications. He took part in World War II in the 18th Armored Division (later renamed the 18th Artillery Division). Thus the world of the soldier and of war is familiar to him not merely from hearsay. Günter Dorn is a talented artist and painter of soldiers, of which there are very few in Germany today. He has succeeded in making the world of the Prussian Army come alive again before our eyes, for the esthetic pleasure of all military scholars, uniform historians and collectors of soldier figures. We wish this book, which in its way can be compared with Menzel's military works, many editions, wide circulation and great popularity.

A. W. Stragand

Editor-in-chief of THE TIN FIGURE

FOREWORD

The Prussian infantry was the backbone of Frederick the Great's army. Their tactical maneuverability, speed and fighting power, achieved through iron training and unattained before that time, created the basis of his astounding success, even though a tenth of their men were unreliable draftees and "foreigners" taken from neighboring countries. Service in the army freed a man from being an underling on a nobleman's estate, uninfluenced by furloughs at home, and so brought about an improvement of status in terms of justice and social position. It was not in vain that Frederick II, after ascending the throne, increased the number of regiments by half from the 32 taken over from his father, all between 1740 and 1743; but since 1713 the army had been the power base of the state that, from scattered possessions, had grown together and grown stronger. His regiments came through the Seven Years' War in victory and defeat. Their next increase, scarcely half as large, did not take place until 1773. For years Prussia remained the land of infantry and was renowned as the "Queen of Weapons". Wherever she set her foot was hers. Thus her steadfastness and her blood tithe played an essential role in the victories of Frederick's army and the rise of Prussia to a great European power.

Since the battle of Mollwitz on April 10, 1741, the Prussian infantry battalions under Field Marshal Curt Christoph von Schwerin (1684-1757) showed ability and fighting power in ruthless attacks, even against numerically greater opponents, or as rocks in the surf with booming fire that "rolled like a constant thunderstorm", according to the report of an Austrian officer. At first a quick attack without fire was preferred. In the infantry regiment of 1743 it was said: "Because the strength of the people and their order makes the Prussian infantry invincible, the people must be enjoined that, if the enemy should stand fast against all reason, their surest advantage will be to charge into them with lowered bayonets, for which the King guarantees that no one will stab back."

In 1761 it was written in "His Majesty the King of Prussia's Instruction on the Art of War to his Generals": "The commanders of the battalions must make sure that the soldier tries to break into the enemy and throw him into retreat, and not to shoot until the enemy turns his back. The battalion's fire does not begin until the enemy is in flight. A battle begun in this way is soon won."

The irreplaceable losses of up to 40 and 50% at Prague and Kolin forced a change, even though the King says in the same "Troop Leadership": "The loss of manpower is not the greatest evil of a lost battle, but rather the fact that the people lose their courage. A general who is beaten must see that through his steadfastness he restores courage to the officer and soldier. He may not make the loss out to be either greater or smaller than it really is."

Led by sixteen drums, the battalion, 100 to 150 meters wide, moved foot by foot, 35 steps per minute, firing from the wings into the ranks, steadily forward. At opponents composed purely of infantry they fired closed battalion salvos of 600 shots, between which the battalion marched forward at 100 steps per minute until they stood right in front of the enemy, one last salvo, then a bayonet attack, supported by two battalions' gunfire in unison, shooting bullets as of 1000 meters and grapeshot as of 400 meters. Six salvos were enough to shake the enemy—other than the Russians. If necessary, the cavalry took care of the rest. Frederick II wrote: "I Ask of heaven that the Prussians never be beaten, and I dare to assert that as long as they are well led and kept in good discipline, they will never need to fear this fate."

The brilliant, if not always exact, drawings of Adolph Menzel have brought Frederick the Great and his soldiers back to life and close to us. The portrayals of Richard and Herbert Knötel, creators of the study of uniforms and considerably influenced by Menzel, have increased the interest. The "History of the Prussian Army" by Curt Jany, laying the groundwork and providing much detail, supported by the work of Dr. Hans Bleckwenn and the evidence of Siegfried Fiedler, have served as a basis for the striking drawings of Günter Dorn of Hannover and the text. It was the publisher's intention to make the high point and most difficult proving period of the infantry of Frederick the Great available to broader circles as an attainable portrayal.

Joachim Engelmann

An infantry regiment consisted of two battalions, each of five companies of musketeers and—since 1735—one company of grenadiers which, in the field, formed one grenadier battalion for every two regiments. The regimental lower staff consisted of a regimental drummer, six oboists, six fifers, gunsmiths and armorers, auditors and provosts, chaplains, quartermasters, a regimental medic with twelve company medics. The mobilizing condition of the manpower was attained by including supernumeraries without weapons, who could then fill vacancies to secure the prescribed three-unit file number. Thus in 1740 it amounted to a service strength of 50 officers, 118 subalterns, twelve oboists and fifers, 38 drummers, twelve carpenters, 196 grenadiers and 1220 musketeers ("old foot"), for a total of 1676 soldiers. In 1743 the King enlarged all grenadier companies by 30 men to 120 privates. In 1756 the King had the old supernumeraries armed and ready for service at the front: thus the musketeer company now had 125 men plus seven without arms, the grenadier company 134 plus six without arms, which means that the musketeer battalion had 702 fighting men and the grenadier battalion 602. With the number of supernumeraries doubled, the regiment had 1320 privates, the grenadier battalion 560 ("middle foot"). In January of 1757 came the next and last strengthening of all companies, again by thirty men. with this "new foot" the regiment had 1620 privates, 1800 soldiers in all, the grenadier battalion 680 privates, 754 men in all. The regiment had almost three battalions in terms of old strength. Only the guards, five regiments without definite reserve areas (Cantons), and the garrison troops still stood on the "old foot", four regiments on the "middle foot", and all others on the "new foot". Reformation was handled in various ways.

Every company had a flag except the grenadier companies, thus there were ten flags in a regiment. King Frederick William I had made their basic designs uniform and fitted them with the royal arms as their standards; only the colors varied from one regiment to another. In 1740 some of Frederick II's generals suggested the motto "Pro deo et patria". The King replaced "pro deo" with "pro gloria", saying: "One must not mix the name of God into human quarrels. War applies to a province, not religion!" The flags were not lowered before the King but saluted by him. They were made of thin silk taffeta, about 1.5 meters square, with the decorations painted on in oil paint. Their durability was not great. Until 1740 they were replaced every five years. But the King commanded: "When the regiments are to have new flags, His Majesty will command when the regiments pass muster." Around 1775 he replaced all infantry flags. The flag of the first company of the regiment was considered the flag of the unit and was made of white cloth with the reflecting surface in color, the opposite of the other company flags. Like uniforms, flags showed a splendid variety in their unity: they were regarded as an honor, their removal as dishonor. The eagle flying against the sun was replaced by the eagle with sword and thunderbolts in its claws. By 1747 all flags had been changed to the king's new pattern.

Shoulder patches to indicate differences in rank existed only from 1808 on, uniformly since 1866. At that time only groups of ranks were differentiated. After many years in rank and file, the men knew each other personally much more closely, so that certain weapons and ornamentations sufficed to make distinctions recognizable. All officers from lieutenant to colonel wore the same uniform in comradely equality, differentiated by age, appearance and medals. The King elevated the generals only in the spring of 1742 by a white ostrich feather along the hat-brim. Officers showed the following insignia of rank: spontoon, developed from the boar-knife in Brandenburg in 1690, still only an honorary decoration; sashes of silver or gold thread, swords with decorated handles and double edges, ring collars of silver with an enamel eagle in the center, formerly protection for the neck; Spanish reed under the arm outside of rank and file, plus metal-colored hat trimmings, held together by a black bow, and embroidery on the coat. Staff officers generally wore riding boots, although on the march all officers from ensign up rode horses. Until 1755 the junior officer could be recognized by his short gun, and until 1713 he still wore a halberd. He had a smooth braid on his hat, likewise stripes on his sleeve-cuffs, a black and white tassel atop his hat, likewise his sword-knot. The sergeant bore a sword-knot on his saber since 1789. Junior officers always wore gloves and had a corporal's staff as their main identifying mark, sometimes hanging from a uniform button. Until 1918 corporal was the only rank of the men. From 1811 on they were allowed to have the junior officer's sword-knot on their sabers.

The infantryman received eight Groschen in pay every five days, two Taler a month. A lieutenant received eleven Taler. Quarters were free, tobacco and bread were cheap, two pounds a day for two Groschen. Whoever was industrious, lived modestly and shared expenses with his comrades could get by very well. Awareness of duty, strict discipline and cleanliness defined the soldier's life. From 1740 on, the infantry used the flintlock musket as its primary weapon instead of the slow-match musket. Rapidity of fire increased, and the use of bayonets made pikes superfluous. The King soon had each man's bullet quota doubled from thirty to sixty, and from 1759 on another sixty bullets per man were carried in the ammunition wagon. The infantryman carried 65 pounds in the field, "closed five times crosswise across the chest". He received a new uniform every year. Since he paid for it, he could dispose of the old one as he pleased. Soon the "blue coat" was worn by many a civilian.

At first the regiments were differentiated as "on foot", "on horse" and dragoons. Since the first foot regiments were only musketeers, this concept became attached to them when, beginning in 1676 in Brandenburg, grenadiers were introduced, following the French pattern, and were formed into companies from 1735 on; the King named his 24 new regiments fusiliers with the exceptions of infantry regiments of I.R. 34 and 50. The traditional "on foot" terminology reappeared in the First to Fifth Guard Regiments on Foot of the Royal Prussian Army in Potsdam, Berlin and Spandau. The fusiliers were smaller soldiers with guns a hand's-breadth shorter, and they wore a sort of smaller-scale grenadier's cap with a spike, while the musketeers wore three-cornered hats. The grenadiers' caps had a semicircular decorated front panel, to which the original tasseled cap with its tassel, stiffened at the rear with whalebone strips, was attached at the top. Otherwise the infantry weapons and uniforms were essentially the same for all three categories. The grenadiers were elite units intended for attacking; from 1740 on, with four companies from two regiments, they were united to form battalions, and they always remained together in the Seven Years' War. The grenadiers of the garrison regiments formed "Constant Grenadier Battalions". Only in peacetime did they return to their original regiments. At that time the regiment was the largest unit of troops. Its commander was always a staff officer, usually a general or a prince whose name was linked with the deeds, fate and renown of the regiment. In the case of similar names, "old" and "young" were applied to differentiate them. The regiments were numbered only as of October 1, 1806, though numbers had been used since 1784 for better differentiation or unification.

The Prussian Army lived from the spirit and example of Frederick the Great, who was a king of the "little people" and practiced justice for all in place of will, haughtiness and a class system. This did not rule out strictness, nor offenses either. It was not customary then to demand more of the officers than of the soldiers in Prussia. In 1773 the King wrote in the "Rules for the Good Battalion Commander in the War": ". . . one knows that the welfare of the troops exists only in the value of their officers . . ." The rebuilding of the gone-to-seed infantry after 1763 is noteworthy. Even in 1778 the King allowed capable soldiers to become junior officers on account of bravery, and good junior officers to become officers and even receive patents of nobility. What the King meant for the army was expressed by Grenadier Hoppe of Anhalt-Köthen:

"Without God and the King, I as a foreigner would not have lasted so long!"

BIBLIOGRAPHY

Geschichte der Preussischen Armee (History of the Prussian Army), by Curt Jany, 2nd ed., Vol. I-IV, Biblio-Verlag, Osnabrück, 1967.

Die geistige Kultur des preussischen Offiziers (The Spiritual Culture of the Prussian Officer), by Friedrich Karl Tharau, von Hase & Koehler Verlag, Mainz, 1968.

Stammliste aller Regimenter und Corps der Königlich-Preussischen Armee für das Jahr 1806 (Register of All Regiments and Corps of the Royal Prussian Army for the Year 1806), new printing 1975, Biblio-Verlag, Osnabrück, 1975.

Friedrich der Grosse (Frederick the Great), Conversations with Catt, ed. Willy Schüssler, Dieterich'sche Verlag Buchhandlung, Wiesbaden.

Die Ritter des Ordens Pour-le-merite (The Knights of the Pour-le-merite Order), ed. Gustaf Lehmann, Vol. I, 1740-1811, Verlag Mittler und Sohn, Berlin, 1913.

Die altpreussische Armee 1714-1806 und ihre Militärkirchenbücher The old Prussian Army 1714-1806 and its Military Hymnals), compiled by Alexander von Lynecker, Verlag Degener und Co., Neustadt an der Aisch, 1980.

Geschichte der Grenadiere Friedrich des Grossen (History of Frederick the Great's Grenadiers), by Siegfried Fiedler, Schild-Verlag, Munich, 1981.

Die Uniformen des Preussischen Infanterie 1753-1786 (The Uniforms of the Prussian Infantry 1753-1786), ed. Dr. Hans Bleckwenn, Part III/Vol. 3, Biblio-Verlag, Osnabrück, 1973.

INFANTRY REGIMENT NO. 1

Regiment on Foot

Commanders of the Regiment

1657		Colonel von Pöllnitz, later Major General
1679		Colonel von Wrangel
1684		Lieutenant General von Schöning, later Field Marshal
1691		Field Marshal von Flemming
1693		Field Marshal von Barfus
1702		Field Marshal Alexander Hermann, Count von Wartensleben, Minister
1723	4/17	Major General Caspar Otto von Glasenapp, later Field Marshal
1742	8/1	Colonel Hans Christoph Friedrich, Count von Hacke, later Lieutenant General
1756	5/27	Lieutenant General Hans Carl von Winterfeldt, died 9/8/1757 at Moys
1758	1/5	Major General Johann Siegmund von Lattorff
1760	2/5	Major General Carl Christoph von Zeuner
1768	2/5	Colonel Ernst Julius von Koschembahr, later Major General
1776	10/20	Colonel Christian Friedrich von Bandemer, later Major General
1778	3/23	Colonel Ludwig Gottlob von Kalckreuth, later Major General
1778	4/30	Major General Hans Ehrentreich von Bornstedt, later Lieutenant General
1792	12/4	Colonel Johann Ernst von Kunheim, since 1798 Count, later Lieutenant General

Out of the "Markish Bodyguard Company" of Captain Wilhelm von Kalckumb, known as Lohausen, of July 21, 1615, and the "Prussian Bodyguard Company" of Captain Pierre de la Cave,of June 1, 1641, there emerged in 1657/1659, with four recruited companies, the "Electoral Guard Regiment," whose commander was Colonel von Pöllnitz, later Major General. His successors were Colonel von Wrangel in 1679, and Lieutenant General von Schöning in 1684. Merged with the 5th Regiment in 1685, it grew to thirty companies by 1688, fought at Warsaw in 1656, Ofen in 1686, and served at Magdeburg in 1663 and Stettin in 1677. From 1689 to 1695 it saw service on the Rhine against the French, and was commanded by Field Marshals von Flemming from 1691 and von Barfus since 1693. In 1702 Field Marshal Count von Wartensleben became the commander of the regiment, of which five companies were transferred to Infantry Regiment 18 in 1703 and ten to Infantry Regiment 5 in 1707. It was now called the "Fusilier Guard" and fought under Prince Eugene from 1704 to 1710. In 1713 four companies were transferred to Infantry Regiment 23 and one to Regiment 24, the Guard rank was given to Infantry Regiment 6, and it was now called the "Wartensleben Regiment on Foot". Scarcely older than Infantry Regiments 2, 3 and 5, it was ranked immediately after the Guards from 1768 on, regardless of its commander's years of service.

On April 10, 1741, under General von Glasenapp, it took part in the Battle of Mollwitz with the Second Battalion. Its Grenadiers in the Kleist Battalion held their position on the right flank in the first encounter despite heavy losses. At Hohenfriedeberg on June 4, 1745 the Ferdinand Brigade, with the right portion of the left wing, made the decisive move to Thomaswaldau against the Austrians who outnumbered it in both manpower and artillery. The regiment defeated three enemy regiments in hard fighting and captured many flags, but lost its commander, Colonel Ewald Wedig von Massow, 17 officers and 631 men. Five Pour-le-merite medals recognized the "quite uncommon bravery of the regiment", whose commander since August 1, 1742 had been Count von Hacke. After the battle it still had 18 officers, 48 non-commissioned officers, 22 musicians and 378 men, and was transferred to Breslau. When on November 23 of that year the King made a surprise attack on the Austrian flank at Hennersdorf (north of Lauban), halting their advance on Bautzen that was intended to cut him off from the Mark, and wiping out four Saxon regiments and taking almost 1100 prisoners, it was again in service under the command of Major General Hans Carl von Winterfeldt.

The regiment did its duty in the Seven Years' War but showed only average accomplishments. From September 10, 1756 on it took part, under its new chief, Lieutenant General von Winterfeldt, who had drawn up the campaign plan and was a highly respected friend of the King after many years of military and diplomatic service, in the siege of the Saxons at Pirna. In Prague on May 6, 1757, Winterfeldt was wounded in the neck at the Homole-Berg in the attack on Sterbohol, while the regiment, north of Kej, joined in the murderous breakthrough of the enemy's center at the high point of the battle, led by Duke Ferdinand von Braunschweig. It lost two-thirds of its strength: 22 officers and 1168 men. Early in September of the same year Winterfeldt led one corps to defend Lusatia; on September 7 it was attacked near Moys by Nadasdy with three times its strength. When the enemy had taken the badly manned Holzberg, he led a counterattack with four battalions and was fatally wounded in the spine. He used his last few hours of life giving inclusive commands like a hero of antiquity. His death was a serious loss for the King and for the defense of the country: "I shall find means to use against my enemies, but I shall find few Winterfeldts," he grieved. Despite its losses, the regiment took part in the Battle of Leuthen and the capture of Breslau at the end of the year.

Though the grenadier companies were still fighting at Zorndorf and Hochkirch in 1758, the regiment turned up again only in 1760 to fight at Liegnitz on August 15 and Torgau on November 3. Here it belonged to Zieten's Corps, the later attack of which saved the victory. The grenadiers stood firm against 4000 men on March 16, 1762 at Gräthen, and in the Battle of Freibergunder Major von Poseck.

In 1806 the regiment belonged to the corps of the Duke of Weimar, later commanded by General von Blücher. After retreating 800 kilometers from Thuringia to Lübeck in four weeks, it surrendered in Ratekau on November 7, 1806, its Third Battalion in Stettin.

The regiment always had its garrison on the New Market in Berlin, its grenadiers in Strausberg. That had been determined by its role and development. Its replacements came from the districts of Storkow and Zauche, in part also from the districts of Teltow and Beeskow and the cities of Storkow, Teltow, Buchholtz, Zossen and Beelitz.

Infanterie-Regiment Nr. 1

INFANTRY REGIMENT NO. 2
Regiment on Foot

Commanders of the Regiment

1656		Prince Boguslaw Radziwill, Governor in Prussia
1668		Colonel Friedrich, Count von Dönhoff, later Lieutenant General
1696		Colonel Otto Magnus, Count von Dönhoff, later Lieutenant General
1717	12/27	Major General Erhard Ernst von Roeder, later Field Marshal
1743	10/30	Major General Samuel von Schlichting, later Lieutenant General
1750	6/12	Major General Hans Wilhelm von Kanitz, later Lieutenant General
1769	11/23	Lieutenant General Joachim Friedrich von Stutterheim
1783	9/22	Lieutenant General Heinrich Wilhelm von Anhalt, Quartermaster General since 1765, Adjutant General since 1770
1786	12/8	Major General Viktor Amadeus, Count Henckel von Donnersmarck
1793	2/9	Lieutenant General Wilhelm Magnus von Brnneck
1805	8/17	Lieutenant General Ernst Friedrich Wilhelm von Rüchel, formerly Guard Regiment Commander

The regiment of eight companies organized in Far Pomerania and the Neumark on December 12, 1655 by Colonel Bogislav von Schwerin was already commanded by Prince Boguslaw Radziwill (1620-1669) in the Battle of Warsaw on July 28-30, 1656. When in 1657 he, as a cousin of the Great Elector, became Governor in Prussia, it remained there in garrison: in Rastenburg and Gerdauen to 1756, Drengfurt, Nordenburg and Angerburg from 1763 to 1765, in Königsberg and Pillau from 1765 on. In 1668 Major General Otto Magnus, Count von Dnhoff took command of the First Battalion, Colonel von Flemming the Second; divided at the time, the two were finally reunited in 1675 and increased to ten companies in 1689. In 1674 they fought in Alsace against Turenne, in 1677 in Pomerania and East Prussia against the Swedes, in 1686 before Ofen and in 1697 at Zenta, in 1705 in the reduction of Cassano, in 1715 before Stralsund, since 1696 under the next Dönhoff. In 1734-35 the regiment took part in the Polish War of Succession on the Rhine. Since 1717 its commander was Major General Erhard Ernst von Roeder. It recruited in the cities of Gumbinnen, Rastenburg, Gerdauen and the districts of Rastenburg, Angerburg, Lötzen, Sperling, Linkuhnen, Kukernese, Heinrichswalde, Wingen, Balgarden, Baubeln and in part Barten, thus being completely East Prussian.

On May 17, 1742 it took part in the Battle of Chotusitz, and in the first half of September 1744 it was active in the encircling and taking of Prague, followed by the laborious return to Silesia via the Sudeten Mountains, beset by hunger, illness and slush. Major General Samuel von Schlichting was its commander since 1743. On February 14, 1745, under Lieutenant General von Lehwaldt, it stormed the Florians-Berg and the Plomitz Heights east of Habelschwerdt, in the process of which it suffered severe losses, including the regimental commander, Colonel Andreas Erhard von Gaudi. At Hohenfriedeberg on June 4 of that year it led the way, with the center of the left wing, in the first line of the second attack left of the gap to the Braunschweig Brigade, which led to victory despite heavy enemy fire. Like all participating infantry regiments, it was allowed to join in the Grenadier March from then on. It was also present at Soor when Schlichting protected the rear of the army from Nadasdy. In four battles against the Russians, 1757-1759, the regiment, whose commander since 1750 had been Major General Wilhelm von Kanitz, suffered heavy losses. After the Russians had moved into East Prussia on August 1, 1757, the 73-year-old Field Marshal von Lehwaldt moved against them south of the Pregel, near Gross Jägersdorf, on August 30, but encountered superior Russian forces on the march in the Norkitten Forest. With lowered bayonets, the regiment broke through the enemy's first line on the right wing and, amid fog, smoke and powder smoke, moved forward bitterly into the second. Without the help of artillery or cavalry, leadership dissolved in the confusion of forest fighting. When the Sydow Regiment mistakenly fired on the rear of the first attack and the Russian counterattack began, the Prussians had to leave their position. The regiment lost 16 officers and 489 men. A year later, almost to the day, on August 25, it and six grenadier battalions led the first attack on the Russians west of Zorndorf in the Zabern valley, followed by the left wing with fifteen battalions under Lieutenant General von Kanitz. In two hours of bitter close combat, it advanced far into the enemy position. But the left wing turned to the right, so that the answering flank attack of the Russian cavalry threw the weakened regiment back toward Zorndorf. Seydlitz's attacks regained some breathing space. When the King began the main attack against the Russian left wing, the advance of masses of enemy cavalry struck the battered regiments of Kanitz. Their second failure in Frederick's eyes brought the East Prussian regiments—with the exception of I.R. 2—into disfavor. The regiment lost 20 officers and 844 men.

On August 25, 1759 Wedell wanted to stop the Russians from crossing the Oder at Crossen, though his forces were only half as strong as theirs. The storming of the Palzig Heights near Kay against the developed battle order of the Russians, with limited artillery support, had to be a bloody failure. Wedell lost a third, the regiment lost 13 officers and 642 men. The King took command himself against the Russians, and on August 12 of the same year—three weeks later—attacked their superior forces in their strongly entrenched position at Kunersdorf, east of Frankfurt/Oder. After good initial success, the regiments came to a standstill in hours of fighting under the August sun at the narrows of the Kuh-Grunde before the Great Spitz-Berg. The Austrian cavalry completed the terrible defeat, which again cost the regiment nine officers and 472 men: 58 officers and 2447 men in two years! The remainder, with the strength of one battalion, was captured by the Austrians on December 4 of the same year after brave resistance at the Elbe bridgehead near Meissen. Since it had lost its own quarters, it went to the Stutterheim Corps in Pomerania to be formed anew with recruits from Mecklenburg. The grenadiers were in action at Torgau on November 3, 1760, led in the third attack by vonHülsen. On August 16, 1762 it experienced the victory of Reichenbach under the Duke of Bevern, and took part in the subsequent siege of Schweidnitz, which fell on October 10.

In 1806-1807 it belonged to the L'Estocq Reserve Corps, which made possible the victory of Preussisch-Eylau on February 7-8, 1807 by its vigorous attack on the birch grove of Kutschitten.

Infanterie-Regiment
Nr. 2

INFANTRY REGIMENT NO. 3

Regiment on Foot

Commanders of the Regiment

1670		Field Marshal Johann Georg, Prince of Anhalt-Dessau
1693 August		Colonel Leopold, Prince of Anhalt-Dessau, later Field Marshal
1747	4/12	Field Marshal Leopold Maximilian, Prince of Anhalt-Dessau
1752	12/23	Colonel Leopold Friedrich Franz, Hereditary Prince of Anhalt-Dessau, later Major General
1758	1/5	Major General Henning Alexander von Kahlden, died 10/22/1758 of wounds at Zorndorf
1759	3/31	Major General Franz Adolf, Prince of Anhalt-Bernburg, later Lieutenant General
1784	4/28	Colonel Heinrich Ernst von Leipziger, later Major General
1788	6/3	Major General Johann Leopold von Thadden
1800	10/1	Major General Johann Jeremias von Renouard

The regiment recruited in Regensburg, Nürnberg and Frankfurt/Main by Colonel Johann von Fargell under a contract of December 16, 1665 was called "Anhalt" since 1679, as Prince Johann Georg von Anhalt-Dessau , Field Marshal, had become its commander in 1670. With its eighteen companies it fought on the Rhine in 1672, in Alsace in 1674, around the isles at the mouth of the Oder in 1675-1676, taking from the Swedes Anklam and Demmin, Stettin in 1677, Rügen in 1678. In 1686 it was in service before Ofen. In 1693 Prince Leopold I of Anhalt-Dessau, the famous "Old Dessauer", took it over from his father. From 1688 to 1695 it saw service in the Netherlands; in the War of the Spanish Succession it saved a victory for Prince Eugene at Höchstädt, in 1705 it suffered the bloody defeat of Cassano, broke through the French left wing at Turin in 1706, and joined in the advance on Toulon in 1707. In 1708 a part of it served in Kirchenstaat, in 1715 it helped to take Stralsund. After that it was at first garrisoned in Magdeburg, from 1717 on in Halle/Saale, with its three grenadier companies in Quedlinburg. Its recruits came from the cities of Halle, Wettin, Löbejün, Cönnern, Leimback and Schraplauas well as the Saale district, the County of Mansfeld and part of the Principality of Halberstadt, and were thus Central Germans. There was often friction between the rawboned drilled regiment and the students at the recently founded university. Other than from 1699 to 1718 it consisted—like the guards—of three battalions, each with a grenadier company and other privileges which ranked it, after I.R. 6, as "the army's best" until 1740. Since 1712 its Chief was a Field Marshal as well as Instructional Master and Drillmaster of the Prussian Army.

In taking the Fortress of Glogau on March 9, 1741 the first two grenadier companies captured eleven flags. At Mollwitz on April 10 of that year the grenadiers under Major von Bolstern stood between the Gendarmes Regiment and the Schulenburg Dragoons on the right wing of the cavalry, meeting the attack of 4000 Austrian cavalrymen. They held "like rocks in the surf, firing forward and backward", and lost 42 dead, 167 wounded. The entire regiment took part with the grenadiers in the conquest of Prague on September 16, 1744. On February 14, 1745 the grenadiers attained their goal by storming the heights east of Habelschwerdt without losing a man. The battle of Hohenfriedeberg on June 4 of that year saw the whole regiment with the grenadiers at the right before the center ofthe first battle line, fighting bitterly without a shot against the Saxons in the wooded lowlands of the "Gule" in resounding, storming battle, with Crown Prince Leopold at their head, losing only 26 men. On September 30 of that year the regiment climbed the fire-spitting Graner Koppe north of Soor with the grenadiers, scaling a slope without shelter. Shortly before the summit a firefight broke out without command; soon two battalions turned and ran. Only the second attack moved the regiment forward again in a bayonet charge, making the heights fall and the outcome clearer. The Battle of Kesseldorf on December 15 of that year was opened by the "Old Dessauer" with his regiment's attack over an ice-covered slope with no cover, three grenadier battalions in the lead; the attack collapsed under fire. To the sound of the Dessau March the regiment renewed its advance but had to give way. Prince Leopold organized the battalions under fire and led the exhausted regiment to break into the Saxon batteries. The counterattack of the Bonin Dragoons decided the matter. The regiment lost a third of its manpower: 14 officers and 519 men.

In the Seven Years' War the regiment faced a crisis as Frederick's relations with the House of Anhalt grew complicated. At Lobositz on October 1, 1756 the King threw the regiment in from right to left as the last reserve to aid the embattled Lobosch after the cavalry fight. At Prague it took part in the right wing's first line under the Duke of Braunschweig-Bevern, breaking through the enemy center south of Kej under the King's leadership. It had many casualties in the infantry battle at the Rokenitz Brook. At Kolin six weeks later, on June 18, 1757 it belonged (without the Third Battalion) to the right infantry wing of the Duke of Bevern, whose avoidance of actionassured a defeat. Then it protected the retreat at Planjan. On November 22 of the same year its grenadiers suffered a defeat at Breslau, on December 5 they experienced a victory at Leuthen. The connection with the House of Anhalt ended with this year; on January 5, 1758 Major General Henning Alexander von Kahlden took command of the regiment. The loss of a battalion at Kolin caused it not to see service again until July 23, 1759, when its three battalions of the avant-garde under Lieutenant General von Manteuffel were defeated at Kay. Despite heavy artillery fire it penetrated to the Russian gun positions, but was thrown back with heavy losses. Four further attacks ended similarly. On December 12 of that year seven companies, without their flags, fell into the hands of the Austrians on the right bank of the Elbe at Meissen. When the King surrounded Dresden on July 10, 1760 the first two battalions were beaten back with heavy losses in the suburb of Pirna, as far as the "Great Garden", during a night attack of the forces in the city on July 21. The King scornfully demoted the battalions to garrison duty, with the loss of their sidearms and the emblems on their coats and hats, which made a deep impression on the army. Three weeks later, after unequaled bravery—"like a corps of furies and devils" they threw back the enemy with lowered bayonets at Bienowitz—they were fully rehabilitated on the battlefield by official proclamation: "The regiment's mark of shame is to be lifted!"

In 1762, under the leadership of its commander, General Major Prince Franz Adolf of Anhalt-Bernburg, it captured Leutmannsdorf, but lost the commander of its Second Battalion. The regiment won two Pour-le-merite medals for its victory.

With the capitulation of Magdeburg on November 8, 1806 the regiment's history came to an end.

Infanterie-Regiment Nr. 3

INFANTRY REGIMENT NO. 4
Regiment on Foot

Commanders of the Regiment

1672		General of the Infantry Christian Albrecht, Count zu Dohna
1677		Major General Johann Albrecht, Count von Barfus, later Field Marshal
1698		Major General Christoph, Count zu Dohna, later General of the Infantry
1716	Feb.	Colonel Jakob von Bechefer, later Lieutenant General
1731	9/21	Colonel Georg Rudolf von Glaubitz, later Lieutenant General
1740	9/7	Colonel Konrad Heinrich von der Groeben, later Major General
1744	7/2	Major General Samuel von Polentz
1745	7/15	Major General Christoph, Count zu Dohna
1745	10/30	Major General Carl Erhard von Kalnein, later Lieutenant General
1757	10/14	Major General Carl Friedrich von Rautter
1758	9/20	Major General Georg Reinhold von Thadden, later Governor in Glatz
1774	6/8	Colonel Friedrich von Pelkowsky, later Major General, Commandant of Koblenz
1782	5/20	Colonel Albrecht Dietrich Gottfried, Baron von Egloffstein, Count in 1786, later Major General
1789	7/12	Major General Benjamin von Amaudruz
1797	5/9	Colonel Wilhelm Heinrich Adolf von Kalckreuth, later Field Marshal

In the spring of 1672 the regiment, with eight companies, was created out of recruits and transfers from the Küstrin garrison for its Governor General of the Infantry, Christian Albrecht, Count zu Dohna. In 1677 it went to Barfus, conqueror of Ofen and Court War Council President in 1696. In 1677 it occupied Stettin. In 1685 it gave troops to I.R. 12, in 1693 to I.R. 17, and in 1700 to I.R. 19. It fought on the Rhine in 1689, one battalion was in Hungary from 1691 to 1695, taking part in the Battle of Szlankamen in 1691, while the other remained before Namur. In the War of the Spanish Succession it was at Oudenarde and Lille in 1708, at Malplaquet and Mons in 1709. It took the Peenemünder fortifications despite heavy losses. Until 1772 it was in Preussisch Holland and Mühlhausen, East Prussia, its grenadiers in Liebstadt, and from 1773 Elbing became its garrison, with the grenadiers in Marienburg. Its recruits came from the cities of Preussisch Holland, Mühlhausen, Gilgenburg and Deutsch Eylau and the districts of Mohrungen and in part Marienwerder and Neidenburg.

Colonel Konrad Heinrich von der Groeben became the regiment's commander in 1740. In the Battle of Chotusitz on May 17, 1742 it took part in the final attack of the right infantry flank with 21 fresh battalions in the second meeting, without noteworthy losses. With that the First Silesian War was over for the regiment. The grenadiers had stayed in camp at Göttin. In 1744 it marched with the King's army to Bohemia under Major General Samuel von Polentz, and took part in the siege of Prague in the first half of September. At Soor on September 30, 1745, the grenadiers under Major Friedrich August von Finck achieved the capture of the Austrian batteries on the Graner Koppe, a key enemy position thought to be impregnable, along with the grenadiers of I.R. 3 and 6. They suffered high losses here: 32 dead and 133 wounded.

The regiment saw remarkably little service in the Seven Years' War. It was now commanded by Lieutenant General C. E. von Kalnein and was in the "Prussian Corps", under the command of Field Marshal von Lehwaldt,that had the task of securing East Prussia, which for an East Prussian regiment meant defending the homeland. The East Prussian troops were mobilized only since February of 1757. The State Militia and the Citizens' Guard in the cities were called upon to take action against bands of robbers and raiders. The Prussians under Field Marshal Count Apraxin crossed the border at Wirballen on August 1, 1757 and united with the corps of Count Fermor west of Insterburg, numbering more than 55,000 men with 79 heavy guns. The 73-year-old Lehwaldt had barely half their strength, with 26,000 men and 20 heavy guns. Nevertheless he hoped he could attack the Russians from the south on August 30 of that year, with the brunt of the attack to the right, and throw them back to the Pregel. His regiments marched bitterly against an enemy who burned the land and committed beastly excesses. Apraxin did not expect an attack and wanted to advance on Allenburg when the Prussians met him at the western edge of the Norkitten Forest. The whole regiment, grenadiers and all, drove the Russians back into the eastern part of the forest with outstanding bravery, losing 78 men and the commander of the grenadier battalion 4/16, Johann Dietrich von Polentz. Isolated during the long struggle in the fire and powder-smoke of the thick woods, the musketeer battalions heavily lost more than twice as the grenadiers, the right flank losing two-fifths of its strength. The growing superiority of the Russians and the lack of leadership finally necessitated an ordered retreat. Apraxin turned at Allenburg and left East Prussia until the end of October, heavily devastated. But the sacrifice had been worthwhile. In the fall of 1757 Major General Carl Friedrich von Rautter became the regiment's chief, one of many changes.

The Russians were already at the Oder in Zorndorf on August 25, 1758, this time led by Fermor and superior to the King's army by 8000 men. The First Battalion belonged to the right flank under Lieutenant General Christoph, Count zu Dohna, Lehwaldt's successor as commander of the "Pomeranian Corps" and chief of the regiment as of 1745. The grenadier battalion under Major von Kleist was part of the left wing, under Lieutenant General von Kanitz, which attacked first. In the morning it stormed the Russian battery against withering grapeshot fire, then it struck the Russian counter-offensive in the flank, and in the afternoon it met the enemy cavalry charge. With 28 dead including its commander, 206 wounded, and 176 missing or captured, it had lost two-thirds of its numbers. At first it was strong enough only to defend a bridge in Schwedt/Oder. When Dohna's flank arrived at the high point, it was hit by the wrath of the Russian main attack led by General Demiku, in which several East Prussian regiments were so badly shaken that they collapsed. Seydlitz's counterattack gave them a breather after what had gone before. The "battalion almost driven into the ground" could not save the regiment from the resentful disfavor of the King toward all East Prussian regiments, and this had its effect even after "a kind of reconciliation" in 1773.

In 1761 Major General von Thadden became chief of the regiment, which had lost its recruitment garrison and thus received only second-rate replacements. Only four years later, on July 21, 1762, was he able to move on Burkersdorf from Ludwigsdorf and storm first the northern battlements and then the southern in the face of a brave defense. Two Pour-le-merite medals on July 26 bore witness to the King's recognition. In 1806 it belonged to the Württemberg Corps, commanded by Blücher, under whom it surrendered at Ratekau and Travemünde on November 7, as did the Third Battalion in 1807 at Danzig under its last regimental commander.

Infanterie-Regiment Nr. 4

INFANTRY REGIMENT NO. 5
Regiment on Foot

Commanders of the Regiment

1655		Colonel Jonas Casimir, Baron zu Eulenburg, later Major General
1667		Prince Boguslaw Radziwill, Governor-General in Prussia
1670		Elector Karl Emil, Colonel, later Major General
1674		Colonel Hans Adam von Schöning, later Lieutenant General
1707		Lieutenant General Georg Abraham von Arnim, later Field Marshal
1731	9/21	Lieutenant General Jakob von Bechefer
1732	2/6	Lieutenant General Christoph Heinrich von der Goltz
1739	4/10	Colonel Johann Heinrich von Wedell, later Major General, died at Chotusitz
1742	7/5	Colonel Anselm Christoph von Bonin, later Lieutenant General
1755	6/7	Lieutenant General Ferdinand, Duke of Braunschweig, later Field Marshal
1766	12/12	Lieutenant General Friedrich Christoph von Saldern
1785	4/6	Lieutenant General Christian August von Lengefeld
1789	8/26	Lieutenant General Ludwig Carl von Kalckstein
1800	11/20	Lieutenant General Franz Kasimir von Kleist

In 1667 Prince Boguslaw Radziwill, as Governor-General in Prussia, took over the Eulenburg Regiment that had been founded in the summer of 1655, filled with enlisted men from the Electoral offices (Wybranzen or "selected ones"). After his death Elector Karl Emil took command and led it against France from 1672 to 1674, until his early death in the Strassburg Hospital. Then Hans Adam von Schöning, who was with it in the field, became its chief; after the capture of Stettin in 1677 and the 1685 campaign against the Swedes and Poles, as Lieutenant General, Governor of Berlin and Privy Councillor for State and War, he united it with I.R. 1 to serve as the Guard. It also fought under him in Hungary in 1686 and on the Rhine in 1689. In 1690 he was discharged "in disfavor" by the Great Elector. It took part in the War of the Spanish Succession under the "Old Dessauer", fighting in Italy from 1705 on, and in 1707 it again became independent as the ten-company "Bodyguard Regiment", serving in the Dauphiné until 1708 and in Flanders in 1709. From 1713 on it bore the name of its longtime chief, Lieutenant General Georg Abraham von Arnim, under whom it saw much service in the Pomeranian campaign of 1715 at Stralsund, before Wismar and in the crossing to Rügen. In 1732 Major General von Bechefer was succeeded as chief by Major General Christoph Heinrich von der Goltz, and with him it went to the Rhine in 1734-1735. Its garrison was Magdeburg. The recruits came from the cities of Magdeburg, Calbe, Stassfurt, Aken, Egeln, Görtzke, Loburg, Luckenwalde and in part from Forest District I of the Duchy of Magdeburg. From 1739 on Colonel Johann von Wedell was a worthy successor to the outstanding chiefs of this elite regiment with its good atmosphere and precise style.

In the First Silesian War only the grenadiers saw action, while the regiment was kept in camp at Göttin as of April 1, 1741 as part of the Old Dessauer's Observation Corps against Saxony and Hannover. The night storming of the Brostau Gate of the Fortress of Glogau was led by Major Hams Carl von Winterfeldt, commander of Grenadier Battalion 5/21, who had just returned from a diplomatic mission to Petersburg to serve as Flank Adjutant to his father-in-law, Field Marshal the Count of Münnich. At Mollwitz four weeks later it was again on the right wing amid the cavalry, where it stood steadfastly and lost 211 men. At Chotusitz—on the right wing again—on May 17, 1742 it suffered scant losses, but the regimental chief, Major General von Wedell, fell. Colonel Anselm Christoph von Bonin became his successor. On June 4, 1745 at Hohenfriedeberg it took part, with the Hertzberg Brigade, in the second line on the left in the last infantry atack of the 18 battalions against 33 Austrian battalions between Thomaswaldau and Günthersdorf. At Soor the grenadiers protected the advance and the rear. Only the regiment took part in the winter battle of Kesselsdorf late that year.

In 1755 Lieutenant General Duke Ferdinand von Braunschweig took command of the regiment. He was the King's brother-in-law and had played a large part in the victories at Hohenfriedeberg and Soor. The entire regiment took part at Lobositz on October 2, 1756, along with the Grenadier Battalion 5/20 under Major Christoph von Billerbeck, leading the midday counterattack from Lobosch to the banks of the Elbe, captured three guns, and lost 98 dead, 14 wounded and 16 missing. Three Pour-le-merite medals testified to the King's approval. The whole regiment fought again as a unit on the left wingin the Battle of Rossbach on November 5, 1757, where the French columns at first attacked in an effort to break through its shallow lines. They were met by the Prussian artillery and the salvos of the regiment, until the attack collapsed. Now the musketeers moved forward, firing, and swung to the right toward the front to the west, while the right wing under Duke Ferdinand of Braunschweig held its ground at Lunstädt. Seydlitz's second attack completed the victory. The losses remained very low, though the regimental commander, Colonel Johann Christoph von Prignitz, was lost. The officer corps received fifteen Pour-le-merite medals as an expression of the King's thanks "because of their particular bravery and good conduct shown in the last battle". At Leuthen four weeks later the musketeers took part, while the grenadiers under Field Marshal von Keithun secured North Bohemia.

In the night attack on Hochkirch on October 14, 1758 the regiment, without its grenadiers, joined the Third Battalion to guard the heights of Pommritz for the right wing under the King, very capably forming a rear line of battle to absorb the troops as they fell back, until the fighting ended about nine o'clock. The grenadiers served at Kunersdorf on August 12, 1759, in the first contact of the avant-garde from the Walck-Berg over the Bächer-Grund to attack on the Mühl-Berg north of Kunersdorf, taking it with more than sixty guns, a very promising initial success that came to a bloody end at the slopes opposite the Kuh-Grund, with 91 dead and 244 wounded. In the Battle of Liegnitz on August 15, 1760 the regiment advanced into the village of Panten during the Prussian counterattack and threw the Austrians back over the Katzbach. The grenadiers did not see action. At Torgau on November 3 of that year, after a long circuitous march, they encountered heavy artillery fire from the enemy center on the Süptitz Heights "in a thin line", suffering heavy losses. The musketeers led the third attack under Hülsen, which failed again. When Zieten's attack from the south began in the twilight, Major von Lestwitz aroused the remains of the regiment to another attack against the surprised enemies, who fell back. The losses were so high that the King forbade any announcement of them. Thanks to its commendable achievements, the regiment always belonged to the King's Army.

In the main army in 1806, it was dissolved in the Magdeburg Capitulation on November 8 of that year; members from east of the Elbe joined the Bodyguard Regiment.

Infanterie-Regiment Nr. 5

INFANTRY REGIMENT NO. 6
Grenadier Guards

Commanders of the Regiment

1675	summer	Elector Friedrich III
1688		Elector Friedrich Wilhelm I
1713		King Friedrich Wilhelm I
1743	3/2	Major General Gottfried Emanuel von Einsiedel, later Lieutenant General
1745	10/27	Colonel Wolf Friedrich von Retzow, later Lieutenant General
1760	2/5	Major General Friedrich Christoph von Saldern, later Lieutenant General
1766	12/20	Colonel Hans Sigismund von Lestwitz, later Major General
1779	6/17	Major General Friedrich Wilhelm von Rohdich, later General of the Infantry, President of the Senior War College
1796	1/29	Major General Adam Dietrich von Roeder
1798	1/19	Colonel Friedrich von Ingersleben
1801	12/21	Colonel Karl Ludwig von Le Coq, Commander-in-chief

When the Swedes attacked Brandenburg in 1674,a call went out through the land to come to the aid of Berlin, and in May and June of 1675 the "Electoral Prince Regiment" on foot was formed for the new Electoral Prince Friedrich (1657-1713), later Elector Friedrich III and King Friedrich I. From 1701 on it was called the "Crown Prince Regiment". As early as 1677 it had taken part in the campaign against the Swedes in Pomerania. In 1685 it provided men for I.R. 12, in 1702 for I.R. 8 and 19. In 1686 it fought at Ofen, in 1689 at Kaiserswerth and Bonn on the Rhine. In 1691 a battalion hired out to the Netherlands covered the retreat at Leuze and beat back two regiments, but suffered severe losses at Steenkerken. In 1695 it fought at Namur, in 1696 at Ghent, in 1697 at Oudenarde. While it besieged Geldern and fought at Höchstädt in 1704, it was strengthened with a third battalion. From 1706 to 1712 it fought in Flanders at Menin, Oudenarde, Ghent, Malplaquet, Tournai, mons, Bethune, Douai, Landrecy and Meurs. Meanwhile Crown Prince Friedrich Wilhelm I formed and paid for the renowned "Great Bodyguard Battalion" in 1710, the "Long Fellows" who were also known as the "Red Battalion" because of their red vests and trousers, and in 1711 he took personal command of the regiment. As is well known, it was not a matter of his whim but rather of the use of long grapeshot rifles. In 1713 the Third Battalion joined I.R. 22, in return for which the Bodyguard Battalion became the First Battalion of the Regiment, now called the "King's Regiment", which had fifteen companies of "big grenadiers" until 1740, and since 1733-1734 three wing grenadier companies, plus a considerable instructional and experimental troop which found many uses for the 25 years of peace after the Pomeranian campaign of 1715. It was closely linked with the Potsdam garrison and drew its replacements—like the I.R. 15—via the "Corps of Unranked Guards" from the field regiments of the army, as well as from the "King's Cantonment" of six Silesian mountain districts, practically an additional battalion. Its sergeants bore the rank of ensign in the army and all its soldiers received higher pay.

On July 28, 1740 Friedrich II retained only one battalion, which as of February/March 1743 bore the name of its chief. Since his I.R.15 now became the "King's Guard", the two other battalions went to join it, the guard company was divided betweeen the new I.R. 34 and 35, and unqualified men were sent to von Weyher's garrison battalion in Magdeburg. Thus the tradition was maintained, playing a recognized though restricted role. At the same time, Major General von Einsiedel became its chief. In the First Silesian War the King left it in camp at Göttin, sending it home in October of 1740. He protected it later as well, as long as possible. At Hohenfriedeberg on June 4, 1745 it took part, with the Guard Brigade of Prince Ferdinand von Braunschweig, in the first line on the left in the last infantry attack, losing 184 men. "Our left wing, after a stubborn fight, in which the regiments suffered particularly . . . along with the grenadier battalion, forced the enemy to retreat via Thomaswaldau and Hohenfriedeberg", it was stated in the granting of three Pour-le-merite medals. At Soor on September 30 of the same year the Guard Brigade broke through the middle of the Saxon and Austrian lines at Burkersdorf, while the Flank Grenadier Company stormed the Graner Koppe with the Tresckow Battalion, losing almost one-third of its men. At the end of October of that year Colonel von Retzow took command of the battalion, and was often recognized by the King in his writings, being called "mon petit Colbert" while Quartermaster General. A decade of peace followed.

In the Seven Years' War it first went into battle at Rossbach on November 5, 1757; the Wing Grenadier Company had already fought in the battles of Lobositz and Prague. In storming the cemetery of Leuthen with the Guard Brigade it achieved "miracles of bravery". This pillar of the Austrian defense was the site of bitter fighting and bloody losses. Here it lost 36 dead and 156 wounded. In the Hochkirch catastrophe of October 14, 1758 it flung itself against the enemy in a closed counterattack of the Guards and Infantry Regiments 20 and 26 west of the village, leaving its left wing a little too far behind. Prince de Ligne wrote: "One simply could not get loose!" In the process it lost about 335 men. In 1760 Major General Friedrich Christoph von Salden became its chief; he was an exemplary teacher of his officers, a great infantry tactician and inspector, a sensible friend to man and father to his soldiers. At Liegnitz it stood at the Katzbach on August 15, 1760, while the left flank and the adjoining Schenkendorffvon Laudon Brigade of Bienowitz were attacked. The losses were slight. In the Battle of Torgau on November 3 of that year, the Saldern Guard Brigade, at the head of the Zieten Corps, struck the enemy on the Süptitz Heights in the rear and flank at twilight via an unoccupied causeway between the sheep ponds, thus deciding the almost-lost battle. The attack cost the battalion eight officers and 338 men. Two years later, on July 20, it took Ohmsdorf Castle, starting point for the attack on the strong Austrian fortifications between Burkersdorf and Leutmannsdorf, in advance of the installation of the great batteries, for the storming, which it then had the job of protecting. At Reichenbach a few weeks later it saw service only after the battle was decided.

In 1779 Major General von Rohdich became Chief of the Battalion. His father, a sergeant in Friedrich Wilhelm I's Bodyguard Company, was held in high regard. He had his son attend the academy in a free position so he could become an officer. Friedrich II always kept him close at hand until his death, and he became Minister of War under the King's successor. In 1801 it was called only the "Guard Grenadier Battalion", was with the main army in 1806 and capitulated in Erfurt and Prenzlau.

Grenadier-Garde
Nr. 6

INFANTRY REGIMENT NO. 7
Regiment on Foot

Commanders of the Regiment

1676	11/27	Electress Dorothea von Brandenburg
1688		Major General Carl Philipp, Margrave of Brandenburg, later Lieutenant General
1695		Major General Christian Ludwig, Margrave of Brandenburg, later Lieutenant General
1734	9/22	Colonel Friedrich Casimir von Botzheim, later Major General
1737	8/30	Colonel Carl Wilhelm von Bredow, later Major General
1741	10/31	Colonel August Wilhelm, Prince of Braunschweig-Bevern, later Duke and General of the Infantry, Governor of Stettin
1781	8/4	Major General Karl Ludwig von Winterfeld
1784	4/5	Major General Wilhelm Heinrich, Baron von der Goltz
1790	5/4	Colonel Karl Eduard von Tiedemann
1790	9/26	Major General Karl Philipp von Owstien, later Lieutenant General

After the Guard Regiment (No. 1) marched off to Alsace in 1674, the four companies that remained at home were expanded to six in the same year and to eight two years later through recruitment, so that on November 27, 1676 the new "Bodyguard Regiment of Electress Dorothea" could be founded. The Electress was (since 1668) the second wife of the Great Elector and came from the house of Holstein-Glücksburg. The commanders were Colonel Gottlieb von Börstel and, from 1687 on, Major General the Count of Schonberg. A year before her death, in 1688, she turned the regiment, now grown to ten companies, over to her son, Major General Carl Philipp, Margrave of Brandenburg, Lord Master of the Knights of St. John, who lost his life on June 23, 1695 after the capture of Casale in Piedmont. In his time the purple-and-red flag bore the Cross of St. John, MC in gold, and a prince's hat above it. From 1677 to 1693 the regiment fought alongside I.R. 6, and it served in northern Italy since 1694. Now it was commanded by his brother, Major General Christian Ludwig, Margrave of Brandenburg, who had just experienced his baptism of fire before Namur with the Spaen Regiment and Prince Leopold of Anhalt-Dessau. Men were transferred to I.R. 12 in 1685, to I.R. 20 in 1688, and to I.R. 19 in 1702; close relations existed with the first and last of these. In the War of the Spanish Succession it fought bravely at Kaiserswerth in 1702, served before Geldern in 1703, fought at Höchstädt and Landau in 1704, then in Italy, at Cassano in 1705 and at Calcinato, Reggio and Turin in 1706. In 1708 it took part in the siege of Lille, and fought at Aire in 1710. It fought before Stralsund in 1715, before a long peace came. In 1734, after the death of his predecessor, Colonel Friedrich Casimir von Botzheim took command, in 1737 Colonel Carl Wilhelm von Bredow, in 1741 Colonel August Wilhelm, Prince of Braunschweig-Bevern, of the reigning line of Braunschweig since 1735, brother of Queen Elisabeth and, according to Berenhorst, "a learned strategist, not always lucky". He wanted "a nice outstanding regiment, save for the guards one of the finest in the army", and kept command until 1781, so that it was linked with the Hohenzollerns for 105 years. Its garrison was in Stettin. Replacements came from the cities of Damm, Poelitz, Bahn, Werben, Freyenwalde, Jakobshagen, Zachan and in part Stettin, as well as from the districts of Pyritz, Randow and in part Saatzig and Greifenhagen. At Mollwitz on April 10, 1741 the grenadiers covered the Ohlau artillery park; the regiment took part in the decisive infantry attack below Schwerin, of which an enemy said: "They marched with the greatest countenance, as if they were on the parade ground." In the Battle of Chotusitz on May 17, 1742 the whole regiment fought together and joined the final Prussian attack of 21 battalions after the cavalries were exhausted and the enemy infantry was held fast in Chotusitz. In September of 1744 the regiment and grenadiers were among the troops besieging Prague. On June 4, 1745 it distinguished itself at Hohenfriedeberg in the left wing's first line, next to the Guard Brigade—with a lot of space between them—by an unbroken attack of Günthersdorf under withering fire, to the extent that Pour-le-merite medals were awarded to "all staff officers and captains". The bloody loss of one-third of its men, among them the Regimental Commander, Colonel Joachim Ernstvon Hobeck, compelled the regiment to stay in Striegau for a time. The grenadiers had come through the storming of Thomaswaldau with light losses. But on September 30 of the same year the breakthrough on the Burkersdorf Heights at Soor cost them 127 soldiers.

On the afternoon of October 1, 1756 the regiment drove back the enemy attack at Lobositz on the Lobosch with bayonets, "at a dead run at full strength with shouts. We went over their dead and drove the enemy back to the Elbe to the right of the village of Welhotta." This attack also pulled the middle and the right wing on to victory. The regiment received eight Pour-le-merite medals, its commander being the "hero" of the day, as also on April 21, 1757 when he opened the path through the mountains to Bohemia near Reichenberg. The regiment did not take part in the Battle of Prague on May 6 of that year, but five weeks later, on June 18, 1757 it was among the leading brigade in the first line and led the advance onto the Krzeczhorz Heights. In the afternoon it stormed an enemy battery "that had done it great damage". After the Prussian cavalry reserve had repulsed the enemy counterattack, it flung itself at the enemy again with bayonets and captured numerous guns. About 5:30 P.M. the enemy line was broken, but the order to back it up with the reserves never reached the Duke of Bevern. When it flew past the Prussian cavalry and, out of ammunition, was attacked from the rear, it went down fighting hand to hand, losing 26 officers and 1018 men! It was possible to restore its numbers only in the following winter. Since November 22 the King scorned the regiment and its chief, whom he blamed for his failure in the Battle of Zorndorf under a massed attack of Russian cavalry on August 25, 1758, saying on January 16, 1774 that it had "disappeared and I did not see it again until the next day. Such a thing can be as little justified as forgotten." On July 23, 1759 it tasted defeat at Kay, on August 12 of that year at Kunersdorf, where the regiment shed its blood before the Great Spitzberg. The grenadiers alone mourned 117 dead and 200 wounded. At Torgau on November 3, 1760 the entire regiment, in the Hülsen columns, twice vainly attacked the Sptitz Heights from the northwest, suffering heavy losses. Only at Freiberg on October 29, 1762 did it capture five cannon and a flag, surrounding the enemy from the south under General von Stutterheim.

In 1806 it was transferred from Weimar's to Blücher's Corps and capitulated at Ratekau on November 7; the Third Battalion remained in Kolberg with all its soldiers in service.

Infanterie-Regiment
Nr. 7

INFANTRY REGIMENT NO. 8
Regiment on Foot

Commanders of the Regiment

1679	2/20	Colonel Johann von Zieten, later Major General
1690		Colonel Anton Günther, Prince of Anhalt-Zerbst, later Lieutenant General
1714	Dec.	Colonel Christian August, Prince of Anhalt-Zerbst, later Field Marshal
1747	3/19	Major General Johann Otto von Treskow
1754	6/25	Major General Georg Friedrich von Amstell, died 5/6/1757 at Prague
1757	5/12	Major General Karl Ferdinand, Baron von Hagen, called Geist, died 2/19/1759 of wounds at Hochkirch
1759	2/25	Major General Julius Dietrich von Queiss
1769	12/13	Colonel Levin Friedrich von Hacke
1785	4/5	Lieutenant General Johann Georg Wilhelm, Baron von Keller
1786	3/1	Major General Johann Andreas Anton von Scholten
1791	8/31	Major General Franz Otto von Pirch
1795	12/26	Major General Friedrich Leopold von Rüts, later Lieutenant General

When the Swedes invaded East Prussia from Lithuania north of Memel on November 15, 1678, only two regiments were stationed there, one newly organized and the other the local conscripts. For that reason a thousand troops from Markish garrisons went to Prussia as a marching regiment to establish a new regiment of eight companies on February 20, 1679, under Colonel Johann von Zieten. In 1688 the First Battalion, commanded by Colonel Anton Günther, Prince of Anhalt-Zerbst since 1690, into the service of the Netherlands, while the Second Battalion, under Colonel Magnus Friedrich von Horn, was dissolved in Brandenburg in 1697. The First Battalion occupied Ath, fought at Leuze in 1691, before Namur in 1692, and won its fame despite losses in the Battle of Steenkerke on August 2-3 of that year and at Neerwinden in 1693, so that the King of England promoted the Prince to Brigadier in 1694. In 1702 it was brought back to full strength with two companies from I.R. 6, stormed Kaiserswerth and Venlo in the War of the Spanish Succession, and fought at Bonn, Maastricht and Huy in 1703. In 1708 it stood before Lille, in 1709 before Tournai, also fighting at Malplaquet, in 1710 it captured Douai, St. Venant and Aire. In 1703 its commander, then a major general, took command of the English-Dutch Corps and was promoted to Lieutenant General in 1705. In 1713 it returned, and after being reorganized into a Prussian regiment it received Colonel Christian August, Prince of Anhalt-Zerbst, as its chief in 1714, entering the war against Sweden as of February 23, 1715, at first strengthening the defense of Stettin, then landing one battalion on Rügen on November 15 of that year. Its garrison was in Minden from 1683 on, in Stettin from 1716 to 1792, in Warsaw and Lowicz for three companies as of 1794. Its replacements came from Stettin, Gülzow, Greifenberg, Labes and the episcopal seat of Cammin, as well as the districts of Flemming, Saatzig, Borck, the Diocese of Kuckelow, and in part Greifenhagen, and since 1795 from the districts of Pultusk, Ostrolenka, Przasnicz, Wyszogrod, Mlawa and in part from Warsaw.

In 1741 the Regiment belonged to the Observation Corps in the Göttin camp, then served in Berlin. Its grenadiers stormed Glogau and fought in the second line of the left wing at Mollwitz. In the Second Silesian War the whole regiment took part in the siege of Prague, where on September 12, 1744 the grenadiers captured positions on the Ziska Mountains. In the subsequent advance into the interior of Bohemia it was among the advance guard under Lieutenant General von Nassau; on October 26 it came through the encounter at Kammerburg during its return to Silesia. On February 14, 1745 the First Battalion, under Lieutenant General von Lehwaldt, stormed the Plomnitz Heights east of Habelschwerdt, liberating the County of Glatz. At Hohenfriedeberg on June 4 of that year the grenadiers were on the left flank, the regiment at the outer left in the second contact, but after the Braunschweig Brigade in the attack, followed by the Bayreuth Dragoons who attacked from there. At the end of that June it again marched to Upper Silesia under Lieutenant General von Nassau, taking Cosel on September 5. The grenadiers secured the rear of the army at Soor and stormed unattackable ice-covered rocky slopes at Kesselsdorf.

In the Seven Years' War it was kept ready for East Prussian service as a Pomeranian elite regiment, but then was sent to Lusatia on December 7, 1756, gaining a victory at Reichenberg on April 21, 1757 under the Duke of Bevern and advancing to join Schwerin's army at Podol. In the bloody Battle of Prague on May 6 of that year the regiment stormed on the right, the grenadiers on the left wing. After Kolin the grenadiers, after bravely defending Gabel to secure connections with Zittau, were captured, but were released the following spring. On November 22 of that year the regiment experienced defeat on the Lohe west of Breslau, but by December 2 had joined the King's army, which came from Rossbach and tried to decide the battle in the renowned maneuver on the plains at Leuthen. In 1758 the regiment again belonged to the King's army, which besieged Olmütz as of June 1 after the fall of Schweidnitz. When the grenadiers had the task of defending a 45-kilometer move back from Troppau successfully at Gundersdorf on June 28 but in vain at Domstadtl on June 30, they suffered heavy losses. On August 25 at Zorndorf they lost another 193 soldiers. At Hochkirch on October 14 of that year the regiment and grenadiers were together under the King. When the great battery of twenty twelve-pounders and six field guns at the southeast corner of the village was lost after bitter fighting, Major General von Geist wanted to reconquer it with his regiment. He was mortally wounded in the process and died on February 19, 1759. The Austrians turned the great battery around and drove off all subsequent attacks with grapeshot, so that the First Battalion, "with leonine courage and under great losses" as well as "with great distinction" could defend only the western edge of town and its gardens. The grenadiers fought on the left wing, which met the second main attack of the enemy. In 1759 it moved with the King's army into the well-defended camp at Schmottseiffen, south of Löwenberg, in order to secure Silesia from the south and the north. In 1760 the regiment was transferred to Prince Heinrich's Corps, which joined the King on August 29 after the victory at Liegnitz, in which the grenadiers took part. In the cold damp November weather at Torgau it took part in the second attack along with Bülow's Brigade after the avant-garde fell back, advancing to the Süptitz Heights and holding off all counterattacks with "greatest bravery", though decimated in a bloodbath. Of 1300 men only 300 came back.

In 1806/1807 it belonged to the L' Estocq Reserve Corps, then to Blücher's Corps in Near Pomerania.

Infanterie-Regiment Nr. 8

INFANTRY REGIMENT NO. 9
Regiment on Foot

Commanders of the Regiment

1680		Major General Johann Ernst von Pöllnitz, Governor of Lippstadt
1684		Major General Heinrich, Baron de Briquemault
1692		Colonel von Boys, died at Neerwinden
1694		Major General von Schlabrendorff, later Lieutenant General and Governor of Küstrin
1713		Major General Etienne du Trossel
1714	Oct.	Lieutenant General Johann Sigismund, Baron vonHeiden, later General of the Infantry, Governor of Wesel
1719	3/14	Colonel Jonas Casimir von Auer, later Major General
1721	3/9	Colonel Johann du Boisson
1726	8/5	Colonel Christoph Rudolf von Schliewitz
1732	1/19	Colonel Sigismund Rudolf von Waldow
1735	7/21	Colonel Otto Friedrich von Leps, later Major General
1747	10/14	Major General Johann Christian Rulemann, Baron Quadt von Wickeradt, died 10/3/1756 after Lobositz
1756	10/4	Major General Friedrich Ludwig von Kleist, died 11/22/1757 at Breslau
1758	1/5	Major General Jrgen Friedrich von Oldenburg
1758	3/1	Major General Nikolaus Lorenz von Puttkamer
1759	7/17	Major General Friedrich August von Schenkendorff
1763	7/31	Major General Carl Friedrich von Wolfersdorff, later Lieutenant General
1782	5/20	Major General Alexander von Budberg
1792	11/7	Colonel Wilhelm Dietrich von Manstein, later Major General
1796	8/18	Colonel Christian Ludwig von Winning, later Major General
1796	12/6	Colonel Johann Friedrich von Brehmer, later Major General
1802	3/20	Colonel Johann Friedrich von Schenck, later Major General

Out of the troops who had been stationed at Minden since 1646 under Colonel von Potthausen, Major General von Kannenberg and Major General von Eller, Major General von Pöllnitz founded the "Pöllnitz Squadron" of 750 men in Lippstadt in 1680, with the inclusion of a free company. Its garrisons until 1756 were Soest, Hamm and Unna; since 1763 one battalion was in Hamm, the other in Soest. Recruits came from Hamm, Unna, Lünen, Bochum, Soest, Altena, Plettenberg, Wetter, Hattingen and the districts of Hamm, Wetter, Altena and Hörde. It served to protect the western parts of the country. At the end of 1694 it was doubled in size with a battalion from the disbanded Alt-Holstein Regiment. Under Major General von Schlabrendorff it was reduced after its return in 1697, being restocked in 1702 by the addition of entire companies from I.R. 3, 10 and 14. It fought before Kaiserswerth, at Rheinbergen and Höchstädt in 1703, before Menin and Turin in 1706, at Oudenarde in 1708 and Malplaquet in 1709. In 1715 one battalion stayed in Magdeburg, while the other took part in crossing the Peene and the Svine and the landing on Rügen. As time passed, its chiefs changed noticeably often.

In the First Silesian War the regiment belonged to the Observation Corps of Prince Leopold von Anhalt-Dessau, which gathered with barely 30,000 men at the Göttin camp south of Brandenburg/Havel at the beginning of April 1741, and was transferred to Grüningen on the Brandenburg-Ziesar road on September 12 of that year "for reasons of health". At the end of October the corps was disbanded. The grenadiers had already gone to the Barony of Herstal/Maas on September 14, 1740, under Major General von Borck, to compel the Bishop of Liege to relinquish the inheritance of the House of Orange. In 1744 the two Westphalian regiments, 9 and 10, were first stationed in Magdeburg. At the end of April 1745 the regiment belonged to the Securing Corps of the Old Dessauer. Mobilized on July 27 of that year, it reached the Dieskau camp near Halle, near the Saxon border, on August 31 and was twice strengthened to 24,000 men. Released on October 15 after tension was reduced, it was mobilized again on November 11 and sent marching to Berlin, but on December 7 it had only reached Torgau. The King angrily called for help; the outcome of the war was in question if the Saxons and Austrians united. After being strengthened by Lehwaldt's Corps, Prince Leopold reached Kesselsdorf on December 15; here the grenadiers on the left flank lost 103 dead and 200 wounded as well as their commander, Major J. E. von Kleist.

While the King rested in his coach on October 1, 1756, the night before Lobositz, surrounded by his generals on the ground, he said over the rattle of gunfire: "That will be the Pandurs; I placed the Quadt Regiment on the hill with them. The Westphalians are rather crude, they strike in the same place three times and throw in a fourth one. But they are good soldiers and will defend themselves." They took the city shortly after 2:00 P.M., after their chief, Lieutenant General Johann Christian Rulemann, Baron Quadt von Wickeradt, had fallen in the artillery fire. In the Battle of Prague on May 6, 1757 the grenadiers fought on the left, the regiment on the right in the first line, under the Duke of Bevern, broke through with the middle south of Kej and entered the narrow valley of the Roketnitz Brook. Here it lost 20 officers and 550 men, half its numbers. Nevertheless, six months later at Rossbach it repulsed the heavy attack of the French, while its chief, Major General Friedrich Ludwig von Kleist, was fatally wounded at Breslau on November 22 of that year and the grenadiers were drawn into the crisis of the left wing at Kolin on June 18. In 1758 the regiment was with the Saxon Corps under Prince Heinrich, the grenadiers were in Upper Silesia. 1759 was a year of destiny: after advancing to Franconia with the Saxon Corps, it was involved in Wedell's defeat at Kay on July 23 and lost its commander, Colonel Johann Gottfried von Kikol. Three weeks later it fought the Russians again at the Battle of Kunersdorf, where its new chief, Major General Friedrich August von Schenckendorff, led the avant-garde, and on November 21, after the heaviest fighting, it was with the rest of the battalion in the surrender of Finck's Corps at Maxen, "an example quite unheard-of to date".

The musketeer Dominicus wrote in his diary: "I don't want to break my oath, but would rather remain true to God and King and bear the burden as long as God wishes"—the testament of a simple soldier!

Only one battalion existed in 1759/1760; there were two again in 1760/1761, with 1220 men despite the lack of replacements from Weestphalia, which was occupied by the French. The grenadiers fought at Torgau in 1760; the regiment was with Prince Heinrich then, with the King again in 1761, seeing service with the Platen Corps above Gostyn in the relief of Kolberg against stronger Russian forces, in unsuccessful forced marches and a retreat through Stettin to Saxony. In 1762 it was seen only in the occupation of Neisse, while the grenadiers fought at Freiberg on October 29.

Its commander since 1763, Major General von Wolfersdorff, had come over from Saxon service only in 1756, had proved himself very well at Torgau and Maxen in 1759, and was an outstanding personality.

In 1806 it capitulated with Rüchel's Corps at Erfurt on October 16, while the Third Battalion surrendered at Hamelin.

Infanterie-Regiment Nr. 9

INFANTRY REGIMENT NO. 10
Regiment on Foot

Commanders of the Regiment

1683	May	Colonel Alexander, Prince of Courland
1686	9/2	Colonel Ferdinand, Prince of Courland
1689		Major General Friedrich, Baron von Heiden, later General of the Infantry
1703		Lieutenant General Friedrich, Crown Prince of Hesse-Kassel, later General of the Cavalry
1714	1/11	Major General Georg, Prince of Hesse-Kassel, later Lieutenant General
1730	5/27	Colonel Dietrich, Prince of Anhalt-Dessau, later Field Marshal and ruling Prince
1750	12/31	Colonel Dietrich Erhard von Knobloch, later Major General
1757	5/12	Major General Gottlob Ernst von Pannwitz
1759	2/10	Major General Friedrich Wilhelm von der Mosel
1768	1/19	Colonel Eggert Christian von Petersdorff
1781	5/22	Colonel Sylvius Ferdinand von Stwolinsky
1787	7/5	Major General Friedrich Wilhelm Siegmund von der Marwitz
1788	12/1	Major General Gisbert Wilhelm, Baron von Romberg, later Lieutenant General and Governor of Stettin
1799	7/10	Major General Wilhelm Albrecht von Burghagen
1799	10/1	Major General Gottlieb Ehrenreich von Besser
1803	12/20	Colonel Karl Alexander von Wedell

In May of 1683 a battalion of four companies was recruited in East Prussia for Prince Alexander of Courland, and enlarged on January 21, 1685 by a second battalion formed from the remaining men of returned marching regiments, becoming the "Courland Regiment" and intended for service against the Turks. At the attack on Ofen on September 2, 1686, which cost it many men and almost all its commanders, the regimental chief also fell, deeply mourned by his uncle, the Great Elector. His brother, Prince Ferdinand of Courland, replaced him. In 1689 he was succeeded by Major General Ferdinand Baron von Heiden, under whom the regiment gave a battalion to I.R. 16 but was refilled with companies from the Spaen Regiment. In 1695 it took part in the siege of Namur. In 1702 some of its men were transferred to I.R. 9. Lieutenant General Crown Prince Friedrich of Hesse-Kassel, later King of Sweden, took command in 1703, and part of the regiment fought in the War of the Spanish Succession in the pay of The Netherlands: in 1706 it was before Menin, at Tournai, Malplaquet and Mons in 1709, it helped Douai and Aire in 1710 and besieged Bouchain in 1711. It was stationed in Brabant in 1712, returning home to be garrisoned in Westphalia in 1713. Now it experienced only successful sieges and investments. In 1714 Major General Prince George of Hesse-Kassel succeeded his brother as its chief. In 1715 one battalion replaced the Schönebeck Battalion in Kolberg; other companies were active in the landing on Rügen on October 15 of that year. Since 1714 Bielefeld and Herford were each the garrison of one battalion. Replacements came from the County of Ravensberg, in part from the Principality of Minden, but mainly from the districts of Hausbergen, Vlotho, Limberg and Sparenberg as well as the cities of Bielefeld, Herford and Vlotho themselves. Twenty-five years of peacetime training followed. Because of the colors of the vests and trousers it was called "The Yellow Regiment".

At Mollwitz on April 10, 1741 the King sent both battalions from the left wing to Crown Prinz Leopold, who led the second battle line, in order to fill the flank with the First Regiment between encounters and extend his forces to the right with the Second. It came through the mass attack of the enemy cavalry and was the first to meet the enemy in the counterattack led by Schwerin, but lost fourteen officers and 439 men, almost a battalion. In December of that year it marched into Moravia with Schwerin's Corps and occupied Olmütz on December 27, then joined the King's army, which advanced to Brno, then marched northward in the spring of 1742. On May 17 of that year it took part in the decisive infantry attack at Chotusitz in the direction of Czaslau, fighting on the right wing. In the Second Silesian War it secured the important fortress of Magdeburg along with I.R. 9, until it became part of the new corps of Prince Leopold of Anhalt-Dessau, formed in April of 1745 on account of the changing attitude of Saxony, and came through the Battle of Kesselsdorf on December 15. Again it lost fourteen officers and 477 men in artillery fire and close combat; the grenadiers lost 72 dead and 206 wounded.

In the Seven Years' War, now under Major General von Knobloch as its chief, it belonged to the King's army—under his command on the right wing of the middle corps—which moved concentrically into Saxony. Unlike the I.R. 9, it did not take part in the Battle of Lobositz. During the Battle of Prague on May 6, 1757 it was part of Keith's Corps, which enclosed the west side of the city with 30,000 men and secured connections to the rear. After Kolin the King lifted the siege of Prague on June 19 of that year, and Keith led his troops back to Leitmeritz. The grenadiers had fought at Kolin and were captured on July 15 after bravely defending the town of Gabel. While the King marched out of the Zittau area to the Saale on August 31, the regiment went to Silesia under the Duke of Braunschweig-Bevern. On September 7 it fought at Moys, when Winterfeldt fell, on November 22 it tasted defeat in the first battle west of Breslau, but fought again at Leuthen on December 5. Along with the Guard, it stormed the enemy's main support point at the cemetery again and again with heavy losses, losing twelve officers and 741 men! The King eulogized them with high praise. Yet it took part in the siege of Breslau, which fell on December 19. In 1758 it marched with the King to Olmütz, to besiege it from June 1 to July 1. The grenadiers secured the move back from Troppau, being attacked on and after June 28 at Gundersdorf and Domstadtl. When the King turned aside toward Zorndorf, it stayed in Silesia under Margrave Karl, and after his departure to join the King it joined Fouqué's Corps, which secured the main army's left wing at Troppau, Neisse, Frankenstein and Landeshut in 1759, while the King was encamped at Schmottseiffen. It was a year that lacked great events but won time! In 1760 Fouqué had had only 12,000 men after transfers, including the First Battalion, while the Second was stationed in Neisse and the grenadiers were with Prince Heinrich's Corps until Freiberg in 1762. The Second Battalion had previously held its own against superior forces at Tepliwoda with great bravery. Fouqué took a fortified position around Landeshut on June 17, was circumvented on the east by Laudon and surrounded by three times his forces. On the rainy night before June 23 the two-sided attack succeeded. "Height after height had to be torn away from the Prussians by bitter counterthrusts, one mountain after another, the entire withdrawal in best order, step by step", Laudon reported. Demands to surrender were refused. The battalion went into captivity with just 328 men. This honorable defeat in battle was compared by the King with the Greeks at Thermopylae! Returned to strength and led by Colonel Count Lottum, it stormed the north slope at Bukersdorf on July 21, 1762 and received three Pour-le-merite medals. With Rüchel's Corps in 1806, it was disbanded at the capitulation of Erfurt on October 16, while the Third Battalion surrendered at Nienburg/Weser.

Infanterie-Regiment Nr. 10

INFANTRY REGIMENT NO. 11
Regiment on Foot

Commanders of the Regiment

1685	8/18	Major General Friedrich Ludwig, Duke of Holstein-Beck, later Field Marshal
1721	8/19	Major General Friedrich Wilhelm, Prince of Holstein-Beck, later Field Marshal
1749	12/4	Major General Lorenz Ludwig von Below, later Lieutenant General
1758	9/12	Major General Johann Carl, Baron von Rebentisch
1763	7/25	Major General Hans von Tettenborn, later Lieutenant General
1776	6/30	Colonel Jakob Rüdiger von Zastrow, later Major General
1782	5/20	Colonel Friedrich Alexander von Rothkirch, later Major General
1786	3/1	Major General Carl von Voss
1790	9/26	Major General Friedrich Carl Ludwig, Duke of Holstein-Beck, later in Russian service
1798	3/1	Colonel Ernst Sigismund von Schöning, later Major General
1807	3/8	Colonel Heinrich, Prince of Prussia, later Major General

On August 18, 1685 Major General Duke Friedrich Ludwig von Holstein-Beck established a free company in the west, which was soon enlarged to a battalion by two companies from the Alt-Holstein and Spaen regiments and became a regiment in 1687, at first garrisoned in Wesel. In 1688 it gave up troops to I.R. 20, in 1702 to I.R. 17. From 1688 to 1697 one battalion fought in the service of The Netherlands against France, seeing service at Namur and Landen, among others; in 1693 one fought against the Turks in Hungary, including at Temesvar and Zenta. In 1705-1706 the grenadiers were stationed in Italy, seeing service at Cassano, Calcinato, Reggio and Turin, while the regiment served in Flanders, for example, before Lille in 1708, at Tournai, Mons and Malplaquet in 1709. From 1710 to 1712 one battalion was still in Italy, while the other served in East Prussia and the Uckermark during the Northern War. In 1715 it joined General von Arnim's Corps at Wollin on May 1, and took part in the landing on Rügen on November 15 with one company in the Fourth Battalion under the command of Colonel Prince Friedrich Wilhelm von Holstein-Beck, then the son of its chief and himself the regiment's chief as of 1721. After its withdrawal from France it was stationed in East Prussia, in Königsberg until 1765 and after that in Rastenburg, Angerburg and Nordenburg, with the grenadiers in Drengfurth. Its recruits came from a third of Königsberg, Angerapp and Drengfurth as well as the districts of eastern Salmland around Labiau-Tapiau as far as the Tilsit-Insterburg Line.

In the First Silesian War it was with the Old Dessauer's Observation Corps, camped at Göttin, from early April to October 1741. As of March 13, 1742 it became part of his newly-formed corps, which was sent marching southward from Jägerndorf at the end of April. When Prince Leopold took command in Upper Silesia at the beginning of May, the regiment went to the King's army, first going into battle at Chotusitz on May 17, grenadiers and all. After the cavalries had fought, the First Battalion went into action in the second line south of the town, while the Second Battalion, called in later, became involved in bitter house-to-house combat under heavy fire, until the attack of the fresh right wing decided the battle. In the Second Silesian War it belonged to General von Marwitz's Second Corps, which secured Upper Silesia as far as Jägerndorf and Troppau. When the King evacuated Ratibor and Troppau and the enemy's 10,000 men threatened Margrave Karl's retreat from Jägerndorf, Zieten rode with five squadrons from Patschkau via Neustadt, which Captain von Ostenreich had bravely defended with 300 men, to Jägerndorf in twenty-two hours, so that the corps, with the regiment and the 300-wagon magazine, could depart for Frankenstein. Attacked at Bratsch, the Württemberg Dragoon Regiment made short work of two Hungarian infantry regiments. On June 4 the regiment closed ranks at Hohenfriedeberg on the right wing in the second line, with the grenadiers on the left flank. As a result of the infantry attack's swing to the left, it came to the fore and struck the gap between Saxon and Austrian troops. The grenadiers, in the Polentz Brigade, took Thomaswaldau. At 9:00 it was all over. At Soor on September 30 of that year they provided connections to the rear. At war's end all East Prussian regiments returned to their garrisons.

In the Seven Years' War it experienced action against the Russians and only one victory. It was in Field Marshal von Lehwaldt's Corps, first seeing action as of August 1, 1757. The battle was lost at Gross-Jägersdorf on August 30 because of confusion in the forest fighting, but the retreat was quite orderly. The grenadiers were almost wiped out. When the Russians withdrew, the King sent the regiment, including the grenadiers, marching to Near Pomerania to meet the Swedes who were landing there; they met early in December, and early in April of 1758 they beat back an attempted Swedish landing at Peenemünde. In January and February of that year the Russians again occupied East Prussia. But the regiment remained with the Pomeranian Corps, now commanded by Count Dohna, in order to hold off the Swedes until the Russians arrived, but it was sent to Frankfurt/Oder late in July, as the Russians had already attacked Küstrin. In the Battle of Zorndorf on August 25 of that year it was temporarily thrown into panic by a massive Russian cavalry attack on the right wing, already under two hours of artillery fire, but was able to win, losing nineteen officers and 707 men and gaining the King's displeasure, although General von Rautter had failed and—along with Major General von Below, the regimental commander—was dismissed immediately. Only at the review of 1773 in Graudenz did the King hold a "special review" of every company and told the officers then: "You know the reason why I was dissatisfied with the regiment". Major von Reibnitz replied: "Your Majesty is right! But I was shot through the body, and many were fatally wounded". The King: "Yes, few of you are still here. If you promise to lead the regiment well at your first opportunity, I'll forget everything. Will you promise that?" They all exclaimed: "Yes, yes, Your Majesty!" The King said: "I won't think about it any more, I'll forget everything".

In 1759 the grenadiers had an easy time at Kay on July 23, and on August 12 they lost 263 men fighting at Kunersdorf. The regiment came through the Battle of Pretzsch successfully on October 29, under its new chief, Major General von Rebentisch, overstepped by Winterfeldt from Austria in 1747, and experienced in secret service and as a leader of troops. After a long fight, it was captured at Maxen on November 20 along with the Finck Corps, as its commander had advised. Restored with difficulty to one battalion in 1760, it defended Breslau from July 31 to August 4, staying there and then taking part in the siege of Schweidnitz, under Tauentzien, from August 8 to October 10.

In 1806 and 1807 it was in L'Estocq's Reserve Corps in East Prussia, and in 1914 it was known as Grenadier Regiment King Friedrich Wilhelm I (2nd East Prussian) No. 3.

Infanterie-Regiment Nr. 11

INFANTRY REGIMENT NO. 12
Regiment on Foot

Commanders of the Regiment

1685	1/23	Margrave Philipp Wilhelm von Brandenburg-Schwedt, later Quartermaster General
1712	3/31	Colonel Heinrich Friedrich, Margrave von Brandenburg-Schwedt, later Major General
1741	5/31	Colonel Wolf Balthasar von Selchow, later Lieutenant General
1743	12/15	Colonel Ludwig IX, Crown Prince of Hesse-Darmstadt, later Lieutenant General
1757	11/10	Major General Friedrich August von Finck, later Lieutenant General
1763	5/29	Major General Johann von Wunsch, later General of the Infantry
1788	10/20	Major General Franz Casimir von Kleist, later Lieutenant General and Governor of Magdeburg
1800	11/20	Colonel Friedrich Wilhelm, Duke of Braunschweig-Oels, later Major General

Acting on a command of January 23, 1685 Colonel Wilhelm von Brandt—Adjutant General in Hungary in 1686, then Governor of Pillau, and commander in Hungary in 1691-1692—organized a 300-man regiment of eight companies at Potsdam on March 16 of that year, out of transfers from the six regiments on foot of 200 men of the "Electress" Regiment, for Margrave Philipp Wilhelm von Brandenburg-Schwedt. The regimental chief was the oldest son of the Great Elector's second marriage, half-brother of King Frederick I, Quartermaster General and reorganizer of the artillery. One year later a battalion took part in the siege of Ofen, beginning on June 18 and ending when Ofen fell to the Brandenburgers on September 2. On account of poor handling, the Supply Corps came back in December and was garrisoned in Lebus, in the Neumark, with its staff in Landsberg. In 1689 one battalion fought at Kaiserswerth and Bonn, in 1690 it was at Vis on the Meuse, in 1691-1692 it was at Mons, Dinant, Liege and Ath and defended Namur, from 1694 to 1697 it took Huy and Namur with one company at first and eight at the end. In 1792 it gave two companies to I.R. 19. As decided by the War Council of December 5 of that year, the regiment's guns and loading procedures became standardized for the entire army, forming the basis of the first Prussian drill regulations of December 18, 1702. After seeing action at Venlo and Roermond in 1702 and being expanded to ten companies in 1703, it came under the command of Prince Leopold of Anhalt-Dessau in 1704 with two battalions and was sent to southern Germany, in order to serve in northern Italy and Provence from 1705 to 1707 and, under von Arnim, to take part in the campaign in the southern French Alps from 1708 to 1712. After its chief's death in 1711, his second son, Margrave Heinrich Friedrich von Brandenburg-Schwedt, took it over but had little success. It returned with 1200 men in January of 1713, after losing about 340. In 1715 one company of the Fourth Battalion took part in the landing on Rügen. From 1716 on it was garrisoned in Prenzlau; its replacements came from the Uckermark and the cities of Prenzlau, Strasburg, Lüchen, and those for the grenadiers from Templin.

The grenadiers opened the First Silesian War and took part in the capture of Ohlau and Troppau. The whole regiment saw service at Mollwitz on April 10, 1741 and came through the crisis unharmed, as if it had been on a parade ground. As a "Prince's Regiment" it seldom saw action. On May 31, 1741 the King took the regiment away from Margrave Heinrich, "because he constantly claimed to be sick", and gave him Fusilier Regiment 42. Colonel von Selchow succeeded him. Under Field Marshal von Schwerin, one battalion occupied Olmütz on December 27 of that year; in 1742 the whole regiment advanced into southern Moravia with the King, but was then sent to Upper Silesia to serve under Prince Leopold von Anhalt-Dessau. In 1744 it joined the King's First Army Corps and moved on Prague, which surrendered on September 16, but then had to endure the return across the Sudeten Mountains early in the winter, in cold, snowy, rainy, stormy weather. On February 14, 1745 it stormed the Plomnitz Heights at Habelschwerdt under Lieutenant General von Lehwaldt, causing the evacuation of the County of Glatz, and it was on the right wing in the second line at Hohenfriedeberg on June 4 of that year. When this developed into a marsh forward, General von Waldburg led his regiments against the Saxon right wing. Then a cannonball knocked him from his horse. General von Bonin gave way to the left under heavy artillery fire, which "was very well executed", and attacked across the swampy meadows and ditches. No sooner was the ammunition used up than the regiment attacked the enemy with their bayonets, pulled the next regiment along with them, and captured seven cannon. Its commander, First Lieutenant von Münchow, fell in the attack. On December 15 it joined the Lehrwaldt Corps in the Battle of Kesselsdorf, where it fought on the right wing.

As early as June 26, 1756 the regiment was mobilized with the Third Prussian Corps in Pomerania, but after the situation changed in December, it was sent to Lusatia to take part in the breakthrough to Bohemia at Reichenberg on April 21, 1757. In the Battle of Prague on May 6 it fought on the right wing in the first line and, under Colonel von Hertzberg, led the breakthrough south of Kej, "at a run over the long causeway at Hostawitz". It was spared Kolin. Its careful training by its chief, Crown Prince Ludwig of Hesse-Darmstadt, who was known as a "militarist", had borne fruit: it was now often put on display nearby. In 1758 it belonged to the Saxon Corps of Prince Heinrich, which was supposed to hold the Imperial Army at a distance. After advancing into the Upper Palatinate, it saw service at Zschopau, Pirna and Dresden from the end of June on, before going to the corps of its commander, Major General von Finck, west of the Elbe. Its grenadiers had fought at Hochkirch. Still in Saxony in 1759, it took part in the second advance into Franconia, and joined the King along with Finck's Corps at Lebus on August 9. At Kunersdorf on August 12 it took part in the opening maneuvers and the attack on the western edge of the Trettin Heights. Superior numbers, terrain and firm Russian steadfastness made a victory impossible. On November 20 it was captured at Maxen. Wunsch, the son of a Swabian master craftsman and free corps leader, had wanted to fight their way out. Reestablished in 1760, one battalion served in Prince Heinrich's Corps, which moved to the Warthe against the Russians, and later in von der Goltz's Corps; in 1761 it fought under Zieten with two battalions, then under Platen at Gostyn on September 15 and under Spie at Kolberg on October 2, serving in Pomerania until year's end. In 1762, with the King's army in Silesia, it captured the southern battlement at Ober-Leutmannsdorf on July 21 with a bayonet charge up a coverless slope. The King rewarded it with two Pour-le-merite.

At last, under the "Black Duke" of the Wars of Liberation, it came to an end with the Weimar Corps on November 7, 1806 with the capitulation of Blücher at Ratekau, while its Third Battalion surrendered in Stettin. The well-trained soldiers in the Uckermark joined Kolberg's Regiment.

Infanterie-Regiment
Nr. 12
PRO GLORIA ET PATRIA
G. Donn

INFANTRY REGIMENT NO. 13
Regiment on Foot

Commanders of the Regiment

1686	Jan.	Colonel Jacques L'Aumonier, Marquis de Varenne, Lieutenant General in 1713
1715	11/27	Lieutenant General Anton von Pannewitz
1722	9/10	Major General Alexander, Count von Dönhoff
1740	6/24	Colonel Friedrich Sebastian Wunibald, Hereditary Steward, Count of Waldburg, died 6/4/1745 at Hohenfriedeberg
1745	7/17	Major General Samuel von Polentz, died 1/28/1746 of wounds at Kesselsdorf
1746	2/12	Major General Philipp Bogislav von Schwerin
1750	11/7	Major General August Friedrich von Itzenplitz, died as Lieutenant General 2/5/1760 of wounds at Kunersdorf
1760	2/8	Major General Friedrich Wilhelm von Syburg, was given I.R. 16, 5/21/1762
1762	6/19	Tsar Peter III of Russia, assassinated 7/17/1762
1763	4/8	Major General Friedrich Wilhelm, Count von Wylich und Lottum
1774	12/18	Colonel Heinrich Gottlob von Braun, Lieutenant General in 1794
1794	12/29	Major General Alexander Wilhelm von Arnim, Lieutenant General in 1806

Eight weeks after the Great Elector's Edict of Potsdam on November 8, 1685, which brought 20,000 Huguenots to Prussia under the assurance of considerable liberty, Colonel Jacques L'Aumonier, Marquis de Varenne, established the sixteen-company 'Varenne Regiment" in Soest. Two years later it was divided to form I.R. 15, the guard regiment of Friedrich II. As a result of frequent transfers to newly-established regiments, it had to be organized anew in 1702, before taking part in the War of the Spanish Succession in the service of The Netherlands until 1713. The regiment had already proved itself to Mollwitz on April 10, 1741 when it stood out particularly at Hohenfriedeberg on June 4, 1745: here it stood in the middle of the first battle line and, with I.R. 19, led the decisive attack on Günthersdorf in the process of the famous swing to the left after the Saxons were beaten. The Bayreuth Dragoons made use of its success for their victorious attacks. But the regimental chief, Lieutenant General the Count of Waldburg, died in action, as did his successor, Major General von Polentz, who died of his wounds at Kesselsdorf. In 1750 Major General August Friedrich von Itzenplitz took over the regiment, refined it and led it in its most important years. On October 1, 1756 it had the job of attacking entrenched Croats in the Lobosch vineyards at Lobositz when the failure of two cavalry attacks and the counterattack of the Austrian cavalry had almost brought the King to give up. It defeated the Croats in hand-to-hand fighting, captured the city and prevented the rescue of the Saxons who were enclosed at Pirna. At Prague on May 6, 1757 the regiment fought on the right wing as Major General von Manstein, after the King had broken through the center of the enemy line, independently led his left wing, which had been losing ground, to attack along with his own brigade. Prince Heinrich led the regiment and I.R. 17 through the steep ravine of the Roketnitz Brook on the northern edge of Mount Tabor, deciding the outcome conclusively north of Hrdlorzez. When the little Prince plunged ahead into the water, the musketeers took him on their shoulders. All resistance collapsed quickly, but the cost of success was high. The regiment did not see action at Kolin. It fought at Rossbach on November 5 of that year, where Seydlitz and the cavalry achieved victory.

At Leuthen on December 5 of that year it was the advance guard in the decisive operation under Major General von Wedell, who led the march off to the right and at 1:00 P.M. attacked the Austrian left wing to open the battle—as "the tip of the sword". The Second Battalion tore Sagschütz away from the Croats, the First Battalion, with Zieten's support, took Gohlaufrom the Württemberg forces. While heavy fighting went on around the Leuthen churchyard, the Prussian right wing attacked the Austrian rear at Saara. In the 1758 campaign the regiment again belonged to the King's army. At Daun's and Laudon's attack on Friedrich's camp at Hochkirch from superior positions in morning fog and semidarkness on October 14 of that year, the King had the regiment attack an Austrian battery southwest of Hochkirch along with I.R. 18 under Prince Franz of Braunschweig, who fell here. Despite grapeshot fire it was able to retake the edge of town, but further advances ended in murderous fire. The shattered battalions had to seek cover in the houses. The regiment lost three-quarters of its manpower, the King one-third of his army. On February 5, 1760 von Itzenplitz died of the wounds he suffered at Kunersdorf. His successor was Major General Friedrich Wilhelm von Syburg, who became regimental chief on February 8, 1760.

At the Battle of Liegnitz on August 15, 1760 against Laudon, the regiment took part without coming into battle. When the King moved out on the circling march from Schilda to the Elbe early on the cold, wet third of November of the same year, in order to have Daun in his high position west of Torgau between Zieten and himself, the regiment was again the avant-garde. It survived being beaten back in the process. The King would have lost heavily if Zieten had not attacked at 6:00 P.M. Since General von Syburg was given command of I.R. 16 on May 21, 1762 and Tsar Peter III made peace with the King, evacuated East Prussia and sent Prussia troops to help against Austria, Friedrich gave him the Order of the Black Eagle and the command of this regiment on June 19, 1762. But on July 17 of that year he was assassinated. The regiment went into battle at Burkersdorf on July 21, 1762; when the King wanted to drive the Austrians out of the Fortress of Schweidnitz in order to take it himself, the Russian forces were ordered back. But Friedrich succeeded in delaying his withdrawal so that he could venture to attack. The enemy had dug itself in strongly on the foothills on both sides of the Weistritz at Burkersdorf and Leutmannsdorf. At the break of day, Wied's Corps struck above Leutmannsdrof to ease the frontal attack at Burkersdorf from the rear. The regiment had to fight its way up a steep slope with no cover along with I.R. 12 and Fusilier Regiment 37, flanked by Croats. The enemy was driven back. Frederick the Great always had a high opinion of the regiment because of its steadfastness. Thus he ranked it with I.R. 1 in priority, just behind his guards. The Huguenot spirit of unity had blended closely with Markish reliability in it, as also in I.R. 15, becoming an example of these valuable qualities to the entire Prussian army. Garrisoned in Berlin since 1724, it drew its soldiers from the Havel area with Glien, Löwenberg and Oberbarnim, as well as the Priegnitz with Friesack, Rhinow, Plauen, Havelberg and the Diocese of Brandenburg. In 1806 it met its end in the capitulation of the First Battalion at Prenzlau on October 28. The Grenadier Battalion under Prince August fought to the last bullet at Bandelow against three French cuirassier regiments. The Second Battalion surrendered at Pinnow on November 4, 1806, as did the Third Battalion at Stettin.

Infanterie-Regiment Nr. 13

INFANTRY REGIMENT NO. 14
Regiment on Foot

Commanders of the Regiment

1626	5/1	Colonel Hillebrand von Kracht
1629		Colonel Steiff von Lauenstein
1632		Colonel Henning von Goetz
1634		Colonel Adam Valtin von Rhedern
1653		Colonel Otto Wilhelm von Podewils
1657		Colonel Pierre de la Cave
1679		Colonel Wolf Christoph, Baron von Waldburg, Hereditary Steward
1688		Colonel Hans Georg von Belling, later Major General
1689		Major General Wilhelm von Brandt, later Lieutenant General, at the time Commander in East Prussia
1702		Colonel Christoph Albrecht von Canitz
1713		Lieutenant General Albrecht Conrad, Count Finck von Finckenstein, later Field Marshal
1736	3/13	Colonel Andreas Joachim von Kleist
1738	8/17	Colonel Johann von Lehwaldt, later Field Marshal
1768	11/23	Colonel Friedrich, Imperial Count of Anhalt
1777	1/11	Major General Johann Christian Wilhelm von Steinwehr
1782	5/20	Major General Viktor Amadeus, Count Henckel von Donnersmarck
1786	12/29	Colonel Franz Ludolf Ferdinand von Wildau
1794	12/29	Major General Karl Leopold von Larisch
1795	7/6	Colonel Georg, Prince of Hohenlohe-Ingelfingen
1803	12/20	Major General Gottlieb Ehrenreich von Besser

Colonel Hillebrand von Kracht mustered a regiment of fifteen companies in Spandau, Frankfurt/Oder, Sommerfeld, Bernau and Fürstenwalde to fight against Wallenstein; it numbered 3000 men on May 1, 1626. Eight of its companies went to East Prussia in 1629; three of them were stationed in Pillau in 1636. As of January 1689 these had grown to eight, then to ten companies under Colonel Hans Georg von Belling. One battalion stayed in Pillau, one went into the war against France and besieged Bonn. In 1689 Major General von Brandt became the commander. In 1691 one company went to Hungary with Dohna's Battalion and fought at Szlankamen and Peterwardein. When three more companies were sent to Hungary, this battalion was separated from the one in Pillau and combined with the Dohna Regiment (I.R. 16). In 1696 it fought in Brabant, in 1698 it returned to Prussia in order—with the secret agreement of the Polish King—to seize Elbing on October 14, 1698, which succeeded only on November 11. In the War of the Spanish Succession it marched to the Rhine under Colonel von Canitz, stormed Bonn in 1703, fought at Höchstädt in 1704, and suffered great losses in the capture of Landau. The Old Dessauer took it along to Italy in 1705 and used it at Cassano and in 1706 at Calcinato and Turin. In 1711 it drove the French from Mont de Vallon near Chaumont. It moved to Pomerania in 1715 and saw action with one company of the Third Battalion in the landing on Rügen. It was once again on the Rhine in 1734-1735.

In 1702 it had had men transferred to I.R. 9. In 1708 King Friedrich I gave the regiment the name "Prince of Orange", which was then the title of his son the Crown Prince, so that Friedrich II became its chief at birth. Friedrich Wilhelm I withdrew the name and the honor in 1713. Since 1719 its garrisons had been Bartenstein for the staff and four companies, Friedland for three companies, Schippenbeil for three companies (since 1751) and Landsberg for the grenadiers until 1770, then Preussisch Holland. Replacements came from the districts south of Königsberg with the cities of Brandenburg, Bartenstein, Preussisch Eylau, Landsberg, Domnau and Zinten. Since 1738 Hans von Lehwaldt had been chief of the regiment, leading it for thirty years and rising from Colonel to Field Marshal!

Not involved in the march into Silesia in 1740, it joined the Dessauer's Observation Corps at Göttin in April of 1741; the corps remained together until the end of October. As of March 13, 1742 it went with this corps to lead the march from Jägerndorf eastward into southern Moravia, marching with the King to the upper Elbe in Bohemia shortly thereafter. On May 17 it took part in the final attack of the right wing at Chotusitz. In 1744 it was in Upper Silesia with the Second Corps, under General von der Marwitz. At Hohenfriedeberg on June 4, 1745 it was at the inner edge of the right wing in the first line. Here it joined the attack against the Austrian left wing at Günthersdorf. At Soor on September 30 it was able to repel the attack of the enemy's left cavalry wing near Burkersdorf, but lost its commander, Colonel Alexander Emil, Count zu Dohna. It returned to East Prussia to spend the years of peace under its chief, now Field Marshal, as Commanding General.

In the fateful year of 1757, with its many scenes of action, the regiment awaited the Russian advance into East Prussia since June 6, and fought on August 31 in the murderous Battle of Gross Jägersdorf, which the 73-year-old Field Marshal von Lehwaldt almost won. With a flag in his hand and with two horses shot from under him, he tried in vain to lead the infantry forward again. After the Russian withdrawal it hastened to Near Pomerania to take on the Swedes. In the Pomeranian Corps under Count zu Dohna in 1758 it moved through Schwedt, Eberswalde and Frankfurt to join the King's Army at Alt-Güstebiese and cross the Oder to meet the Russians. In the two-day Battle of Zorndorf it was on the right wing, which the King held back at first, but which was then struck by the Russian cavalry attack yet still gained the victory. After the battle it remained encamped with the corps at Blumenberg, across from the Russians.

On July 23, 1759 it fought under Wedell in the bloody defeat at Kej, where it was unable to wrest the Palzig Heights from the Russians. Three weeks later, on August 12, it fought at Kunersdorf in Finck's Corps, which led the attack that turned southward toward the Trettin Heights. It fought very bravely at the Mühlberg and the Kuh-Grund, losing 35 officers and 717 men! Early in September it secured Saxony against superior enemy forces. Moved forward to Maxen by the King to break the allies' connections to the rear, Finck's Corps was compelled to surrender to three times its strength in the snowy mountains on November 20. The King's displeasure lasted beyond the reconciliation of 1774; the Count of Anhalt went into Saxon and Russian service in 1776 because the King remained irreconcilable. What remained of the regiment was sent to Pomerania by the King to join Stutterheim's Corps and be restocked with recruits from Mecklenburg. Under the Duke of Württemberg it defended Berlin from October 5 to 9 and then spent the winter in Mecklenburg. From early June to mid-December 1761 it fought at Kolberg and took part in the Battle of Freiberg, Saxony in 1762 under Prince Heinrich.

With the First Battalion in East Prussia with L' Estocq's Corps, the Second in Graudenz and the Third in Danzig, it remained in existence and joined the Grenadier Regiment King Frederick the Great (3rd East Prussian) No. 4, thus carrying its critic's name into the future.

Infanterie-Regiment
Nr. 14
PRO GLORIA ET PATRIA
G. Dorn

GUARD REGIMENT NO. 15
First Battalion Bodyguard

Commanders of the Regiment

1688	8/1	Colonel Carl Philipp, Baron von Wylich zu Lottum, later Count and Field Marshal
1719	3/14	Colonel Friedrich Wilhelm, Baron von Koenen
1720	Sept.	Colonel Christoph Heinrich, Baron von der Goltz, later Major General
1732	2/29	Colonel Friedrich II, Crown Prince of Prussia, later His Majesty the King Commanders of the First Battalion:
1740	6/23	Colonel Friedrich Wilhelm, Margrave of Brandenburg, later Major General
1744	Dec.	Major General Ferdinand, Prince of Braunschweig
1755	6/7	Major General Johann Ludwig von Ingersleben, died 11/27/1757 at Breslau
1757	11/27	Colonel Bogislav Friedrich von Tauentzien, later General and Count of Wittenberg
1763	6/25	Colonel Friedrich Wilhelm, Prince of Prussia, later King Friedrich Wilhelm II
1764	4/24	Colonel Hans Christoph von Billerbeck
1765	11/22	Colonel Otto Heinrich von Laxdehnen, later Major General
1773	2/6	Major Ernst Gottlob von Scheelen, later Major General
1786	8/10	Major Franz Georg von Kunitzky
1793	1/9	Lieutenant Colonel Friedrich von Ingersleben
1798	1/19	Colonel Karl Ferdinand von Hirschfeld, later Major General

After Elector Friedrich III, later King Friedrich I, came to the throne, a regiment was founded by command on August 1, 1688, composed of eight companies of the thus halved Varenne Regiment no. 13 and transfers from I.R. 11, commanded by Colonel Carl Philipp, Baron von Wylich zu Lottum, later Senior Court Marshal and Field Marshal, organized in Wesel and having refugees from France as one-third of its manpower. In 1689 the First Battalion's five companies were sent to the Rhine, in 1691 they went to Flanders under Derfflinger, served in Brabant under Fleming in 1691-1692 and fighting at Ghent, Ath and Namur. The Second Battalion went to Hungary under Lieutenant General von Barfus and took part successfully in the battles of Szlankamen, Peterwardein and Zenta. From 1693 to 1697 they were led by Lieutenant General von Brandt and included two companies each from Wesel and Halberstadt. In 1695 the regiment was first reunited and then transferred to I.R. 17. In 1693 the First Battalion fought in the Battle of Neerswinden under Lieutenant General von Heiden, remained in Huy in 1695 and came home with six companies in November of1697. Since February 1, 1699 it was again a full regiment of ten companies. When the War of the Spanish Succession began, one battalion remained in Minden while the other went to war under von Heiden and returned in the fall of 1702 to Wesel, where General von Lottum was Governor. When he became the commander of the Prussian Corps in 1703, the whole regiment went with him, captured Rheinberg and Geldern, fought at Höchstädt in 1704, captured Landau and Menin, the latter in 1706. At Oudenarde in 1708 it fought on the right flank and won a shaky victory for the allies. After serving at Winnendael it fought particularly well at Malplaquet in 1709 and besieged Ghent, Bruges and Ath. In 1715 it stayed in the west at first and saw no further action.

When Field Marshal Count von Lottum died in 1718, Colonel Baron von Koenen became the regiment's chief on March 14, 1719, to be succeeded a year later by Colonel von Goltz, under whom it was transferred to Prignitz in the Kurmark. After his pardon in November of 1730, Crown Prince Friedrich II was allowed to rejoin the army and wear the uniform of the regiment before, on February 29, 1732—after Major General von der Goltz had been named chief of I.R. 5 and Commandant at Magedburg on February 6—becoming its chief himself and transferring it to Ruppin and Nauen the next year. He wrote on October 23 of that year: "I dedicate myself to the regiment," and in the spring of 1733: "I come from drill, I drill, I shall drill". Lieutenant Colonel von Bredow was appointed by the King to advise him; from 1734 on he was joined by Colonel von Camas and Major Senning, who were actually his military teachers. When he came to the throne, he named it his "Guard Regiment" on June 1, 1740. In 1735 King Friedrich Wilhelm I had praised the regiment, embraced his son publicly and promoted him to Major General. Even though he had substantially given up the actual leadership of the regiment after the autumn of 1736, he still took his command seriously and felt responsible for his regiment.

Then there was a reorganization: on August 4, 1740 the First Battalion in Ruppin became the "Bodyguard", a guard and escort battalion of the King, as well as an exemplary and instructional troop, above all for tactical changes. Foreign officers were also trained in peacetime, and after 1763 Prussian staff officers were given thorough training as well. The Second Battalion in Nauen, with its five companies, the half Guard Company I.R. 6, and recruits formed the basis of I.R. 34 for Prince Ferdinand of Prussia, the King's youngest brother. In return, the former Second Battalion of Guard Regiment No. 6 took its place on August 3 of that year. The previous battalions had each had twenty officers, 55 non-commissioned officers, eighteen musicians and 540 musketeers. There was also a Third Battalion formed of men transferred from the army in exchange for men of the I.R. 6. The King himself commanded the First Battalion, the Second and Third—completely separate in terms of organization and equipment—had a joint regimental commander. "Palace and House Guards" as, for example, in France, were unknown in Prussia. The men were, to be sure, called grenadiers, but only the Third Battalion wore grenadier caps. The sergeants bore sword-knots, and those of the First Battalion were equal to first lieutenants, those of the Second and Third to ensigns in the army. The officers were two steps in rank above those of the army, had the finest uniforms and white feather decorations on their hats like generals.

Utilized on the right wing at Mollwitz on April 10, 1741, the regiment held its position very steadfastly. The King "also wanted to thank them for the victory after the battle". Afterward only 180 of 800 men were fit for duty; the losses added up to sixteen officers and 424 men, more than half! At Kolin on June 18, 1757 it followed in the second battle line and finally survived the falling back of the decimated first wave. After that Colonel von Tauentzien led the battalion energetically against the enemy without any support. The enemy fell back under his heavy fire but attacked again with five squadrons of dragoons and infantry when the fact that the battalion was alone became apparent. Constantly surrounded and in grapeshot fire, it fought its way free three times until it had only 300 men. Twenty-four officers and 475 men were lost. From 1758 on it remained in Breslau and fought bitterly under General von Tauentzien to defend the city in 1760. At Liegnitz on August 15 of the same year it protected the baggage train from Austrian hussars at Kuchelberg.

Despite the capitulation of Prenzlau on October 28, 1806, which included the Grenadier Company, it was allowed to form the Guard and Second Company of the new Guard Regiment in 1808, later the First Guard Regiment on Foot.

Infanterie - Regiment
Nr. 15, I. Btl.

GUARD REGIMENT NO. 15
Second and Third Batallions

Commanders of the Regiment

1740	6/23	Colonel Asmus Ehrenreich von Bredow
1743	10/23	Colonel Felix Bogislav von Schwerin, died 6/7/1745 after Hohenfriedeberg
1745	6/11	Colonel Hans Caspar Ernst von Schultze, died as Lieutenant General 11/22/1757
1747	4/22	Colonel Dietrich Richard von Meyerinck
1749	8/24	Lieutenant Colonel Christoph Moritz von Beschwitz
1754	3/3	Colonel Samuel Christoph von Meseberg
1756	7/18	Colonel Carl Ferdinand, Baron von Hagen, called Geist, died as Major General at Hochkirch
1757	5/12	Colonel Friedrich Christoph, Baron von Saldern, later Lieutenant General
1760	2/5	Major Wichard Joachim Heinrich von Möllendorff, later Field Marshal
1771	5/7	Colonel Julius Treusch von Buttlar
1776	10/20	Colonel Friedrich Wilhelm von Rohdich, later Lieutenant General
1779	6/17	Colonel Wilhelm Magnus von Brnning
1785	4/5	Lieutenant Colonel Friedrich Adam Dietrich von Röder
1796	1/29	Colonel Friedrich Wilhelm Felix von Schwerin, later Major General
1800	9/9	Colonel Ludolf August Friedrich von Alvensleben
1805	1/5	Major Johann Ferdinand von Franckenberg Commander-in-chief:
1798	1/13	Lieutenant General Ernst Friedrich Wilhelm von Rüchel, until 1805

Naturally, the regiment was located at Potsdam until 1740. It had no replacement region because it was to consist of a personally chosen quality elite, the "finest team in the whole army". Thus every infantry regiment annually sent three men, every cavalry regiment two, often chosen by the King himself, to the training and replacement battalion of the regiment, the "Corps of the Unranked" (i.e., the not yet assigned). From 1750 to 1763 ten recruits a year came from the districts of Löwenberg-Bunzlau, Hirschberg, Landeshut-Bolkenhain, Jauer, Schweidnitz, Reichenbach—exempt from any draft by virtue of being weavers—to be added to the Guards. There was also direct individual recruitment. A contemporary reports: "Drill was a party, they looked forward to maneuvers. When they were not standing in rank and file, they were uncontrolled, for no one but the King and their officers was allowed to command them".

The Guard Regiment, formerly "King's Regiment", with its Second and Third Battalions was an instructional troop except in war, where it was an elite fighting force. Every battalion had its own commander, always a Lieutenant Colonel and a Colonel, under a common regimental commander. Soldiers and non-commissioned officers were better paid than the regular army. It went with the newly organized Third Army Corps on the march to Silesia in 1741. Only the First Battalion was present at the Battle of Mollwitz on April 10, 1741. Marshal Belle-Isle, staying in camp afterward, observed a drill of the Second Battalion and reported to the French court his admiration of its splendid appearance as well as its training and discipline. At the beginning of 1742 the regiment, with both grenadier companies, was in North Moravia and Upper Silesia, but then marched together with the King to northern Bohemia in order to fight in the Battle of Chotusitz on May 17, taking part in the decisive attack of twenty-one battalions. The short peace between the wars was—according to the King—"a continuing school of warfare. Everyone worked with vigor toward the highest development of every discipline that made the Romans victorious over everyone else".

In 1744 the regiment belonged to the First Army Corps, which moved into Bohemia and besieged Prague. In the process of its winter withdrawal across the Elbe, the enemy formed a bridgehead at Telschitz on November 19. The grenadier battalion of the flank companies of I.R. 15 and 18, under Lieutenant Colonel Georg Vivigens von Wedell flung itself against the bridgehead and held it alone for several hours despite the loss of 184 men. The King compared this achievement with that of "Leonidas". At Hohenfriedeberg on June 4, 1745 the same battalion attacked the Saxon right wing from the Gule without a shot and lost 166 men, while the regiment, in the Guard Brigade of Prince Ferdinand of Braunschweig, fought on the inner left wing, leading the attack of eighteen battalions in line at Günthersdorf against twice their number in the Austrian left wing, and losing the commander of the Third Battalion, Colonel Felix Bogislav von Schwerin. All the staff officers and important participants received the Pour-le-merite. At Soor on Deptember 30 the Grenadier Battalion lost 310 men on the Graner Koppe; among the 1289 dead was its commander, von Wedell. When the attack was halted by artillery fire at Burkersdorf after a cavalry charge, Prince Ferdinand rallied it and the Second Battalion forward and broke through the enemy center with bayonets. Here the regiment lost fifteen officers and 389 men. In 1756 the regiment, in the King's army, surrounded Pirna. At Lobositz only the Grenadier Company of the First Battalion saw service. In 1757 it took part in the siege of Prague and received two Pour-le-merite for holding off the attack of the occupying forces on May 23-24. The commander of the grenadiers, Lieutenant Colonel Daniel Gottlob von Bülow, fell there on June 20. At the end of August it marched to the Saale with the King—while the grenadiers remained in Breslau—and took part in the Battle of Rossbach on November 5, in which only seven battalions of the outer left wing came into battle. After the sixteen-day forced march from Leipzig to Parchwitz to join forces with Zieten, the King took on the enemy at Leuthen on December 5: "The two battalions attacked the village, chased out the Austrians and took seven cannon". One superior officer hesitated before the bulwarks of the cemetery, all around which fighting went on. Then Captain von Möllendorf shouted: "A different man forward!" and stormily burst through the blown-up gate on the east side. "After the village was past, they found a new line of grenadiers and advanced on them. They had no more bullets but kept on advancing and threw the grenadiers for a loss". One battalion commander, Lieutenant Colonel von Diericke, fell, as did seventeen officers and 501 men. Six Pour-le-merite! At Hochkirch on October 14, 1758 the Guard could not change the army's bad luck: the Third Battalion was able to form a parrying line on the Pommritz Heights, but the grenadiers on the left wing were taken prisoner, and thirteen officers and 618 men fell, most of them from the Second Battalion. After Liegnitz in 1760—here the Third Battalion helped to hold off Laudon's attack at Panten—Möllendorff led the regiment, in Zieten's Corps, over the causeway between the sheep ponds at Torgau to attack the Süptitz Heights, which cost the Third Battalion heavy casualties but helped to win the battle. Seven Pour-le-merite expressed the King's gratitude. At Burkersdorf the regiment, under Major General von Möllendorff, stormed the battlements in a dangerous frontal attack on July 19, 1762.

Lieutenant Colonel Daniel Gottlob von Bülow, fell there on June 20. At the end of August it marched to the Saale with the King—while the grenadiers remained in Breslau—and took part in the Battle of Rossbach on November 5, in which only seven battalions of the outer left wing came into battle. After the sixteen-day forced march from Leipzig to Parchwitz to join forces with Zieten, the King took on the enemy at Leuthen on December 5: "The two battalions attacked the village, chased out the Austrians and took seven cannon". One superior officer hesitated before the bulwarks of the cemetery, all around which fighting went on. Then Captain von Möllendorf shouted: "A different man forward!" and stormily burst through the blown-up gate on the east side. "After the village was past, they found a new line of grenadiers and advanced on them. They had no more bullets but kept on advancing and threw the grenadiers for a loss". One battalion commander, Lieutenant Colonel von Diericke, fell, as did seventeen officers and 501 men. Six Pour-le-

Infanterie-Regiment
Nr. 15

II. Btl. III. Btl.

merite! At Hochkirch on October 14, 1758 the Guard could not change the army's bad luck: the Third Battalion was able to form a parrying line on the Pommritz Heights, but the grenadiers on the left wing were taken prisoner, and thirteen officers and 618 men fell, most of them from the Second Battalion. After Liegnitz in 1760—here the Third Battalion helped to hold off Laudon's attack at Panten—Möllendorff led the regiment, in Zieten's Corps, over the causeway between the sheep ponds at Torgau to attack the Süptitz Heights, which cost the Third Battalion heavy casualties but helped to win the battle. Seven Pour-le-merite expressed the King's gratitude. At Burkersdorf the regiment, under Major General von Möllendorff, stormed the battlements in a dangerous frontal attack on July 19, 1762.

On October 28, 1806 the regiment, under Hohenlohe, surrendered at Prenzlau. Out of its remaining men King Friedrich Wilhelm III organized the new Guard Regiment in 1808, including the Fifth to Eighth Companies, later the First Guard Regiment on Foot.

THE UNRANKED GUARDS

The Foot and Horse Guards needed their own replacement and training battalion, not only because of their particular needs in terms of replacements but above all on account of their function as a guarding and instructional troop, for parades, as an experimental group and for service in all representations that were necessary for the King and the army at the time. For all of this not only the normal training of the field regiments was required, but so was additional training, beyond a doubt. Then too, in exceptional cases unqualified people could be sent back at the right time. Hints and details of the history and development of this battalion are few, for this Fifth Battalion of the Foot Guard was not very interesting. Bleckwenn refers to the relationship between the elite troops and the instructional unit in regard to the Reichswehr, whereby one must bear in mind that the Reichswehr units combined twelve different annual classes and had problems as a result.

The "Red Bodyguard Battalion" organized by Crown Prince Friedrich Wilhelm I in 1710 were referred to as "unranked" as early as 1712, though their numbers were small. After 1713 the unit gradually grew to be a "Corps" with the strength of a company, and was so called when transferred to the First I.R. 6 on June 1, 1716 and to the First I.R. 15 on June 1, 1740., still remaining the King's "Guard". It was not to be exchanged with the most highly trained of the field troop units. It was always associated with the First Battalion of the Guard, and at times also called the "Second Bodyguard Company". In the 1722-1723 fiscal year Jany commanded one sergeant, eight non-commissioned officers, six drummers and 160 grenadiers, an increase of 34 soldiers. In 1726 its numbers already totaled 214 . Eight years later it was under the command of officers and included three officers, thirty non-commissioned officers, eight drummers and 300 grenadiers, a total of 341 soldiers. The grenadiers were now divided into three groups: great, second and third unranked. A year later there were almost 400 soldiers; their numbers amounted two about one-seventh of the entire regiment.

The garrison was naturally in Potsdam. When the regiment marched out in 1741, a hundred unranked stayed behind at Nauen out of more than 400 soldiers. Between 1746 and 1756 the "unranked", without a firm budget, usually numbered at most three officers, ten to fifteen non-commissioned officers, ten or more Junkers (cadets), seven to eight drummers and 300 to 400 privates, barely the equal of one battalion, with constant personnel changes. After the terrible losses at Kolin it was completely merged with the First Guard Battalion and only reestablished after the war's end in 1763. At first its numbers remained smaller than before the wars; for example, its strength in June of 1771 was ten non-commissioned officers, three drummers, 83 privates, and in 1778 one officer, nine non-commissioned officers, one medic, 139 privates, a total of 150 soldiers. Its composition was very probably left to chance. It was also a question of the peacetime budget, with the Guards' numbers being reduced from 1792 on and the Corps dissolved in 1806.

A list of commanders is unfortunately not available. The uniform was like that of the Bodyguard Battalion but, without any decorations, more suitable for training purposes, resembling those of the garrison troops and the field uniforms. The actual recruits are said to have worn loose blue linen clothing before being put in uniform. Since one remained a guard soldier for life in Prussia, invalids were also members, and in 1730 their own "Corps of the Guard Invalids" had been formed. The invalids of the Bodyguard Battalion remained with the troops, while those of the Guard Regiment went to that corps, which was garrisoned at Werder near Potsdam. They were called on for guard duty at the castles and in Berlin. Their commander was originally a captain, later a sergeant of the guard with a captain's rank.

It is interesting that as of January 1, 1719 Prince Leopold of Anhalt-Dessau likewise organized a third battalion for his regiment, I.R. 3, and as of June 1, 1720 founded a "Corps of Young People" as a replacement troop, with three non-commissioned officers, two drummers, one piper and sixty men, "in order to draw future recruits for the King's Regiment from it". It was located in Dessau and Oranienbaum, as Jany reports. Very clearly, the far-reaching instructional concept of Pietism and the Enlightenment had worked its way significantly into military life, advanced by King Friedrich I, who had attracted Leibniz, Thomasius and August Hermenn Francke, with the Prussian State University at Halle, founded in 1694, as the focal point.

Infanterie-Regiment Nr. 15
Corps der Unrangirten

INFANTRY REGIMENT NO. 16
Regiment on Foot

Commanders of the Regiment

1689	3/1	Colonel Alexander, Count zu Dohna, later Field Marshal
1738	7/13	Colonel Adam Christoph von Flanss, later Field Marshal
1748	7/14	Major General Christoph, Count zu Dohna, later General of the Infantry
1762	6/19	Major General Friedrich Wilhelm von Syburg
1771	5/7	Major General Christian Ernst Wilhelm Benedikt von Borck
1771	11/1	Major General Ludwig von Buddenbrock
1782	5/20	Major General Karl Adolf Schott von Schottenstein
1785	4/5	Colonel Johann Friedrich Wilhelm Moritz, Baron von Romberg, later Major General
1789	2/7	Colonel Ernst Heinrich von Gillern, later Major General
1792	5/20	Colonel Friedrich Wilhelm Heinrich von Hausen, later Major General
1799	10/1	Colonel Otto Friedrich von Dierícke, later MajorGeneral

According to a contract of March 1-11, 1689, Colonel Alexander, Count zu Dohna (1661-1728), cousin of the subsequent king, established the First Battalion with two companies of the Westphalian Courland Regiment (No. 10), one company of the Briquemault Battalion (No. 9), one company from Ansbach and one company of transfers from Prussian garrisons. Its commander for thirty-nine years later served for eight years as instructor of the Electoral and Crown Prince Friedrich Wilhelm I, chief of the Prussian State Ministry in Königsberg, President of the Prussian Chamber, Field Marshal and Governor of the Fortress of Pillau. On October 12, 1689 it took part in the storming of Bonn. During that winter it was enlarged with three more companies of the Courland Regiment, and early in July of 1690 six companies went to the Rhine, Maas and Mosel while two companies stayed in East Prussia and were joined by two more in 1691. In that same year three companies, strengthened by one company each of Waldburg and Brandt and known as the Dohna Battalion, went to Hungary and fought successfully at Peterwardein and Szlankamen. On their return in 1692 the Dohna companies went to the Brandt Regiment in exchange for three from Pillau, since Count Dohna was the governor there. Now five companies each were at Pillau and in the area of Kaiserswerth, Rheinberg, Wesel and Lippstadt; in 1693 they marched to Liege, and in 1695 they helped to capture Namur after the battalion commander, Colonel Paul Thévenin des Glaireaux, had fallen. Their march back to Pillau took place only in November of 1697; the Second Battalion was still independent and gave one company to I.R. 17 in 1702. In June of 1705 the First Battalion had to go to the Palatinate, where it took Hagenau on October 6. After spending the winter in central Germany, it fought in June of 1706 before Menin, in 1708 at Oudenarde, Lille and Ghent, in 1709 at Malplaquet and in 1710 before Douay and Arras. Still in the field, it was united with the Pillau garrison battalion on June 1, 1713, and returned to Mörs only that December, later being sent to East Prussia. In 1715 it protected this menaced land. In 1734-1735 it belonged to the supply corps on the upper Rhine. The many connections of the Count zu Dohna to the House of Orange in terms of kinship, military service and administration find expression in the color of the flag. In 1721 it was garrisoned in Königsberg, in 1796 it moved to Braunsberg with three companies going to Mhlhausen and the grenadiers to Preussisch Holland. Its recruits came from Angerburg, Lötzen, Rhein, Barten, one-third of Königsberg and the surrounding districts.

In 1741 the regiment was encamped at Göttin and joined the King's army in Bohemia only in April of 1742, commanded by Prince Leopold of Anhalt-Dessau. The Battle of Chotusitz on May 17, 1742 was its first under King Friedrich II, who forced a victory within two hours with the unused right wing of the infantry. In 1744 it belonged to von der Marwitz's Second Corps serving in Upper Silesia. At the end of the years it went back to Neisse, but was recalled by the Old Dessauer early in 1745, advancing to Jägerndorf and to the Oder between Ratibor and Kosel, and seeing action in numerous battles until that autumn to clear Upper Silesia on the right side of the Oder of Hungarian forces. From August 26 to September 5 it cooperated in recapturing the Fortress of Kosel. The grenadiers fought at Hohenfriedeberg and Soor.

In 1756 it was with the Third Prussian Corps under Field Marshal von Lehwaldt in East Prussia, the corps having been assembled at Insterburg on June 6, 1757 to take on the Russians. At Gross Jägersdorf on August 30, 1757 it fought "very well" on the left wing under the leadership of Lieutenant General Count Dohna, its chief and the brother of its founder. The attempted surrounding, which promised success, ended in failure through the confusion of the forest battle. After the Russian withdrawal it attacked the Swedes in Pomerania. At the end of Marsh 1758 Count Dohna was given command of the corps which was to attack the Russians if they returned. In mid-August they crossed the Warthe. The King united his army with Dohna's corps, crossed the Oder and took on the Russians at Zorndorf on August 25. Here it fought very bravely in the first line and lost 21 officers and 611 men, almost a battalion. The King's lingering displeasure—whether because the East Prussian congress swore allegiance to Tsarina Elisabeth on January 24, 1758 or because of the East Prussian regiments' panic at Zorndorf—also affected this regiment. Never again did the King set foot in East Prussia. In 1782 he said to the regimental commander, General von Buddenbrock, whom he valued highly: "Your regiment did a bloody bad job in the last war". He forbade any care for the invalids of the regiment. After his army left, it remained camped with the corps at Blumberg, until on October 29 the King moved from Stargard to Near Pomerania. On the Warthe from mid-June to July 20, 1759, it was sent against the Russians on the Palzig Heights at Kay by Lieutenant General von Wedell; the result was a bloody defeat. Kunersdorf followed on August 12; in the infantry battle between the Kuh-Berg and the Great Spitzberg it lost sixteen officers and 550 men without achieving anything, but at the end of October the Russians withdrew. There were no more replacements from East Prussia.

In 1760 it fought with the Stutterheim Corps, then with the Duke of Württemberg, securing Berlin from October 4 to 8, then moving southwest and meeting the King. In the Battle of Torgau on November 3 it marched behind the cavalry, forming the reserve under Kleist, and led the last counterattack in the twilight when Zieten's attack began. It was able to decide the victory in the last hour without losses. After being in winter quarters at Mecklenburg, it took part in the fighting around Kolberg in Pomerania during 1761 and stormed the battlements at Spie on December 12. In 1762 it returned to Silesia and captured six cannon at Leutmannsdorf on July 21.

Stationed in Danzig in 1806-1807 and then in East Prussia, it was able to join the King Friedrich I Grenadier Regiment (Fourth East Prussian) No. 5 in 1808.

Infanterie-Regiment Nr. 16

INFANTRY REGIMENT NO. 17
Regiment on Foot

Commanders of the Regiment

1693		Colonel Baltzer Friedrich von Sydow, later Major General
1703		Colonel Friedrich Wilhelm von Grumbkow, later Field Marshal
1739	8/17	Colonel Ernst August de la Chevallerie, Baron dela Motte
1748	11/4	Colonel Adam Friedrich von Jeetze, later Lieutenant General
1756	7/20	Major General Gerd Heinrich von Manteuffel, later Lieutenant General
1764	5/23	Major General Carl Gustav, Baron von Rosen
1772	1/16	Major General Constantin von Billerbeck, later Lieutenant General
1786	3/1	Major General Heinrich Gottlieb von Kenitz
1786	12/4	Major General Wilhelm Magnus von Brünneck, later Governor of Königsberg
1793	2/9	Lieutenant General Carl Albrecht von Räumer
1795	7/6	Colonel Carl Friedrich von Lange, later Major General
1802	3/20	Major General Carl Peter von Tresckow

The regiment founded in 1693 and made of the Second Battalion of the Lottum Regiment (No. 15) and one company of I.R. 4 was made independent on March 14, 1695 under Colonel Baltzer Friedrich von Sydow after it had proved itself in Hungary in 1694 and at Peterwardein in 1795. It also took part in Prince Eugene's victory over the Turks at Zenta on September 11, 1697. In 1702 in entered the service of The Netherlands as a full regiment, fighting at Höchstädt in 1703, on the Moselle in 1705, at Oudenarde in 1708, Malplaquet and Mons in 1709, and before Douai in 1710. Reorganized as a Prussian regiment after its return in 1714, it was garrisoned in Köslin and Rügenwalde, where it remained since 1739 and had the districts of Stolp, Schlawe, Gollnow, part of the Principality of Cammin, and the cities of Stolp, Rügenwalde, Gollnow and Köslin as its recruiting area. In the fall of 1715 it cooperated in the conquest of Stralsund against Charles XII of Sweden. Then came twenty-five years of peace.

After marching into Silesia on December 16, 1740, it took part with one battalion in the Battle of Chotusitz on May 17, 1742. At Hohenfriedeberg on June 4, 1745 it was on the right flank next to the cavalry in the second encounter and moved northward at the head of the attack while pushing back the Saxons. On September 30 of that year at Soor, where the King, in his own words, "deserved to be beaten if the bravery of my troops had not protected me", the regiment bore the brunt in the storming of the fire-spitting Graner Koppe in the second line after the attack had been turned back. Here the regimental commander, Colonel Georg Joachim von Blanckenburg, fell. The losses numbered one-fourth of its manpower. Even so, it fought successfully at Habelschwerdt and Neustadt, where it guarded the withdrawal. Eight pour-le-merite, all the way down to company commander, showed appreciation for the steadfast Pomeranian regiment. On November 4, 1748 Colonel Adam Friedrich von Jeetze succeeded the former regimental chief, Lieutenant General Ernst August de la Chevallerie, Baron de la Motte.

From 1756 on the regiment was not so successful, but brave in adversity. On July 20, 1756 Major General Gerd Heinrich von Manteuffel took command of the regiment, remaining its chief until May 23, 1764. At Lobositz on October 2, 1756 the regiment exchanged fire with the Croats by the Lobosch for five hours, until they ran out of ammunition. When the Austrian counter-attack began about 1:00 P.M., it went into action with bayonets and clubs of its own accord, took Welhotta, and threw nine fresh battalions back in the city. In the process it suffered the most losses. "Never have my troops worked such wonders of bravery", the King wrote. On May 6, 1757 the regiment fought in the infantry battle at Prague, serving on the right wing between the Hlaupetin Ridge and the ravine of the Roketnitz Brook. Six weeks later at the Battle of Kolin, the regiment was to march off to the left after the King's breakthrough on the Krzeczhorz Heights, in order to take the area once and for all. The command did not reach them. Despite the disintegration of Friedrich's battle plan, it stood fast for thirty minutes at Chozenitz with I.R. 20 under General von Manstein, in artillery fire and under attack, keeping the retreat open for the beaten left wing. On September 7 of that year at Moys it met the Austrian attack under Count Nadasdy east of the Neisse and lost General von Winterfeldt in the counterattack. When on November 22 of that year the Austrians drove the weak army of the Duke of Bevern that was protecting Silesia, back from the Weistritz toward Breslau while the King was on his way from Rossnach, the regiment, with its last battalion, was in the midst of the defeat. At Leuthen—as at Rossbach and later at Zorndorf—only the grenadier battalion was involved: "It held on very bravely and suffered greatly". Only on March 15, 1760 did it distinguish itself again when, underLieutenant General Baron von der Goltz, it had the job of escorting a magazine from Neustadt O/S to Steinau and was attacked by Laudon with 3000 cavalry, 2000 Croatians and 1900 grenadiers and twice refused to surrender. For several hours "they commenced to hack at us six times, but did not succeed once", von der Goltz reported, having lost only 35 dead, 43 captured and 79 injured, while the enemy lost 800 men. The King expressed his thanks with six Pour-le-merite and wrote: "I hope that other generals take a good example from this action, for this is the old Prussian way of defending oneself successfully even against vast enemy superiority. Convey compliments to Manteuffel's officers in my name. They have acted in our manner, with honor, and not accoprding to infamous modern examples". Laudon said he had not received any deserters "because the regiment consists of native-born Pomeranians". At Torgau on November 3 of that year, under General von Bülow, it moved against the Zinna Heights in the second main attack, captured four cannon, but suffered the bloody loss of 1000 men, half its strength. In 1761 it served with Prince Heinrich in Saxony. There it fought at Freiberg on October 29, 1762 under Major General von Alt-Stutterheim, pushing the enemy right wing back to the south along the Spittelwald "with clever utilization of the terrain".

On October 17, 1806 the First and Second Battalions of Cröllwitz were pushed back to the Saale by Bernadotte while marching toward Halle. Two Junkers drowned rescuing their flags. The remainder surrendered in Magdeburg. The Third Battalion, after an honorable capitulation in Danzig, joined I.R. 2.

Infanterie-Regiment
Nr. 17

INFANTRY REGIMENT NO. 18
Regiment on Foot

Commanders of the Regiment

1698	Nov.	Colonel Anton von Pannewitz, later Lieutenant General
1703		Colonel Daniel von Tettau, died at Malplaquet as Major General
1709		Major General David Gottlieb von Gersdorff
1731	11/1	Colonel Georg Vollrath von Kröcher
1738	6/15	Colonel Christian Reinhold von Derschau, later Major General
1742	11/10	Colonel August Wilhelm, Prince of Prussia, later General of the Infantry
1758	12/8	Colonel Friedrich Wilhelm, Prince of Prussia, later King Friedrich Wilhelm II
1790	3/28	Colonel Friedrich Wilhelm, Crown Prince of Prussia, later King Friedrich Wilhelm III
1797	12/21	His Majesty King Friedrich Wilhelm III of Prussia

Out of transfers from the six guard battalions, Colonel Anton von Pannewitz, later to command I.R. 13, organized a "Grenadier Guard" battalion in Berlin in November of 1698. It served in the wars in the North, West and South under Colonel Daniel von Tettau and was expanded to a regiment in 1703 by the addition of one battalion of the "Prussian Guard". Several of its companies existed even longer. Tettau, by then a Major General, fell at Malplaquet in 1709. In mid-June of 1705 it went to the Moselle with Arnim's Corps and stormed Hagenau on October 6 when the occupying troops broke out. It came home in November and was halved in May of 1713, when six companies were transferred to I.R. 23 under Major General von Kamecke, later Forcade, and called the "White Grenadier Battalion"—to compete with the "Red Bodyguard Battalion", or just to reduce the guards under Friedrich Wilhelm I? After the Pomeranian campaign, while it was doing guard duty in Berlin, it got its Second Battalion back through the induction of six hundred Swedish prisoners from Stralsund and once more became a field regiment under Major General David Gottlieb von Gersdorff: "The fellows are dissatisfied. And when they took off their grenadier caps in the King's presence and received hats, some of them threw them on the ground with curses. They have always been called castle cavaliers. The officers have had a hard time calming them down and have even said that all the men laughed at them, to which the officers said that since the situation was new, it ought to be allowed of the people. They will soon get these notions out of their heads." Now Bernau, Strausberg and Köpenick were the garrisons, and until 1724 the staff was quartered in Frankfurt/Oder. From 1736 on, Alt-Landsberg and Oranienburg were also included. After the Seven Years' War the King moved it to Potsdam to keep an eye on it. The recruits came from the garrisons themselves, plus Lenzen, Werder, Wittenberg, the Altmark as far as Tangermünde and the Prignitz from Wittenberge to Arendsee.

On December 15, 1740 it joined the King's Army Corps at Crossen to march on Breslau. It was not involved in the Battle of Mollwitz, but took an active part in Crown Prince Leopold's advance toward Jungbunzlau at the end of October, 1741 and was supported by the Schwerin Corps at Olmütz at the end of December. Its chief, Major General von Derschau, accomplished the capitulation of Glatz on January 9, 1742. Early in February the First Battalion went with the grenadiers to join the King at Olmütz, then served under Derschau in Bohemia, though without fighting at Chotusitz. In the Second Silesian War it belonged to the King's First Army Corps; the grenadiers, in Wedell's Battalion at Telschitz on November 19, flung themselves unselfishly against the enemy bridgehead. During the march back to Silesia the King strengthened the weak occupying forces in Prague with the regiment. In breaking out and fighting its way through, it lost 747 soldiers to debility, hunger and cold. Then it served in 1745—as did other regiments—in occupying the Fortress of Glogau. When the Old Dessauer assembled a corps at Gattersleben as of August 21 to fight against Saxony, Prince Dietrich provided him with troops, including the regiment. At Kesselsdorf on December 15 it served on the left wing and carried the last attack over the Kessel Brook onto the ice-covered Pennrich Heights , including close combat with bayonets. Since 1742 its chief was Major General August Wilhelm, Prince of Prussia, the King's oldest brother, designated as his successor since 1744 and "seriously loved" by the King, but somewhat weak. In 1756 the regiment belonged to the King's Army, helped to surround Pirna, break through at Reichenberg on April 21, 1757 and fight at Prague on May 6. On the right wing in the first line under the Duke of Bevern, it led the penetration south of Kej against the enemy center, storming over the causeway south of Hostawitz and fighting in close combat by the Rokenitz Brook. On May 9 it took the Ziska-Berg but lost Colonel von Strantz. After Kolin, when the King sharply criticized the withdrawal to Bautzen on July27 because of his brother, he left the army and died unreconciled at Oranienburg on June 12, 1758. On November 22 the regiment came through the defeat at Breslau but lost Lieutenant Colonel Christoph Wilhelm von Belling. One battalion fought at Leuthen on December 5, while the regiment besieged Breslau, In 1758—after a ten-day forced march from Landeshut to Zorndorf with the King's Corps—it fought on the right wing in the crisis at Zorndorf on August 25, putting down the stubborn Russian defense. Before the battle the King had said while reviewing them: "My friends! Stand firm, all will go well!" Four Pour-le-merite, one of them for Lieutenant von der Hagen, recognized this "miracle of bravery". On October 14 it suffered very heavy losses in the fighting around Hochkirch and led the counter-attack on the right wing, preventing a total catastrophe. With the King again in 1759 as "my dear brother's regiment", it was spared the defeat at Kunersdorf. In 1760 it took part in the siege of Dresden in July, did not see action at Liegnitz on August 15, led the last attack at Torgau on November 3 with the Second Battalion in the Saldern Brigade at Zieten, and won five Pour-le-merite for its company commanders. A hundred volunteers under Captain von der Hagen broke through the entanglements and battlements of Burkersdorf on July 21, 1762, while the regiment encircled the guard to the left. Finally, it took part in the fighting at Reichenbach.

In the War of the Bavarian Succession the regiment, commanded by Colonel von Winterfeld, was surrounded in Neustadt/Oberschlesien by 12,000 men late in February of 1779 and compelled to surrender. But it defended the city, which had been set afire by gunfire. Since 1763 it rewarded with privileges because it belonged "to the ruling house", and on April 24,1764 it was named "Prince of Prussia". Not in vain did its grenadiers bear the black eagle to the fore. On December 9, 1786 it became the "Regiment of Prussia", on March 28, 1790 the "Crown Prince Regiment" for Colonel Friedrich Wilhelm III, and after his accession of the throne on December 21, 1797 it was again "the King's Regiment" with guard status. In 1806 it was dissolved with the grenadiers by the Prenzlau Capitulation of October 28, while the Third Battalion was in Spandau; those who remained joined the Guards.

Infanterie-Regiment Nr. 18

INFANTRY REGIMENT NO. 19
Regiment on Foot

Commanders of the Regiment

1702	2/14	Lieutenant General Albrecht Friedrich, Margrave of Brandenburg-Schwedt, later General of the Infantry
1731	6/24	Colonel Carl Friedrich Albrecht, Margrave of Brandenburg-Schwedt, later General of the Infantry, died 6/22/1762 of wounds
1763	4/8	Major General Hans von Tettenborn, given I.R. 11
1763	6/25	Lieutenant General Friedrich August, Prince of Braunschweig
1794	3/20	Major General Karl Ludwig von Gtze
1806	2/17	Lieutenant General Wilhelm Friedrich, Prince of Orange-Fulda, later Prince of Orange

For the half-brother of King Friedrich I a regiment of twelve companies, with 648 privates, was organized as of February 14, 1702, made up of two companies each from I.R. 4, 6, 7 and 12 as well as transfers from the garrisons of Kstrin, Driesen, Spandau, Peitz and Frankfurt as well as the free companies of Marwitz, Börstel and la Cave. Its chief was Lieutenant General Albrecht Friedrich, Margrave of Brandenburg-Schwedt, Commander of the Prussian Supply Corps in 1702-1703, Lord Master of the Knights of St. John from 1696 to 1731. In 1703 it gave two companies to I.R. 10; in 1702 the First Battalion went westward and fought at Kaiserswerth, Venloo and Roermond. In 1703 the regiment was stationed at Geldern, in 1704 on the Lower Rhine and Maas. While the First Battalion remained and served at Ostend, Menin and Ath in 1706, in Brabant in 1707, at Oudenarde, Hondschoote and Mons in 1708, the Second Battalion went to Italy in 1706: Calcinato and Turin in that year, Toulon and Susa in 1707, returning in 1708. In 1709 the entire regiment very bravely penetrated the French line at Malplaquet on September 11, fought at Douay and, one battalion only, at Aire in 1710, and at Bouchain in 1711. The Second Battalion came home then; the First only at the end of 1712. In 1713 one battalion took part in the occupation of Stettin on September 29 while the other served on the Peepe Line until January of 1714. In 1715 the regiment, under Colonel G. E. von der Lieppen, stormed the battlements at Peenemünde on August 22, landed on Rügen with one company in the Fourth Battalion on November 18, and helped to capture Stralsund on December 23.

Until 1737 it was garrisoned at Königsberg/Neumark with the smaller towns of the area, then the Friedrich-Stadt district of Berlin. Its recruits came from Küstrin, Driesen, Königsbert/N, Treuenbrietzen and the districts of Königsberg/N, Soldin, Arnswalde, Friedeberg, Sternberg and Dramburg. After his father's death, the uncle of Friedrich II, Colonel Carl Friedrich Albrecht, Margrave of Brandenburg-Schwedt, 1705-1762, the most capable of this line, succeeded as commander of this "house regiment". He was Lord Master of the Knights of St. John from 1731 to 1762 and highly esteemed by the King for his reliability, bravery, insight and concern. Being connected with the Order for sixty years, the regiment bore the Cross of St. John on its flags, grenadier caps, ring collars, pocket panels, drummers' cords and drum rings, all of which passed to I.R. 34 in 1763 and was replaced by a simple cross.

Part of the Duke of Holstein-Beck's Second corps in 1740-1741, it stormed the Fortress of Glogau in one hour during the night of March 9, 1741, along with two other regiments. Four weeks later it fought at Mollwitz, suffering considerable losses, including Colonel Friedrich, Margrave of Brandenburg, its commander. In the fall it went to northern Bohemia; in 1742 it secured the Fortress of Glatz. In the King's First Corps in 1744, it took part in the capture of Prague and endured the winter withdrawal over the Sudeten Mountains into Silesia. Margrave Carl secured Upper Silesia in the spring of 1745. At Hohenfriedeberg on June 4 it was on the inner right wing in the first line while Margrave Carl commanded the two nearest brigades to the left, and it closed the gap between itself and the Guards in the main attack against the Austrians at Günthersdorf. Four months later it came under heavy artillery fire at Soor-Burkersdorf. On its second approach its commander, Colonel Konrad Gottfried von Buntsch, was killed, as were many others. The grenadiers took part in the attack on Kath. Hennersdorf on November 23.

In 1756 the regiment helped to encircle Pirna under the leadership of its chief. In the Battle of Prague on May 6, 1757 it was on the right flank in the second encounter, breaking through south of Kej and fighting at close quarters in the valley of the Rokenitz Brook. It was spared Kolin. Afterward the King turned westward at the end of August to defend his inner line. At Rossbach on November 5 it was in service with its grenadiers, without losses. The political situation had changed. On December 5 the whole regiment attacked the churchyard at Leuthen, "under small arms fire from beginning to end and almost completely ruined". It also besieged Breslau, which fell on December 19. The King praised its having "worked wonders" but gave no Pour-le-merite. In 1758, after the capture of Schweidnitz and the siege of Olmütz, it remained in camp at Plagwitz under Margrave Carl until it joined the King on September 11. In the night attack on Hochkirch on October 14 the grenadiers fought bitterly to defend the great battery at Rodewitz, losing 209 dead and 50 prisoners; the First Battalion, at the big battery at the southeast corner of the town, and the Second, under Major von Lange and Lieutenant von der Marwitz, at the churchyard, fought for hours against attack by superior forces without giving in. The dead and dying piled up. After using up their ammunition, the last men broke out of the back gate in a bayonet charge, during which von Lange collapsed with eleven wounds, dying several days later. The enemy accorded him the highest honors. In the avant-garde at Kunersdorf on August 12, 1759 the First Battalion and grenadiers stormed the Russian batteries on the Spitzberg in vain, losing 276 men, and were shot down in the Kuh-Grund, while the Second Battalion, on the right flank before the Kudenberg, "almost went to ruin". The survivors came through the Battle of Korbitz on September 21, led by Finck, fighting "as the Prussian Infantry is accustomed to do". In 1760 it proved itself at Strehla as at Wittenberg, where it had to suffer additional losses. At Torgau on November 3 the Second Battalion served with the Queiss Brigade under the King, the First with Zieten, suffering heavy losses.

Despite its recognition as the nucleus of his army, to which the regiment always belonged, and its high esteem in the army, the King still criticized it in 1774 because of Kunersdorf: "Except for Lange's Battalion, (I had) no reason to be satisfied with the regiment, which did not want to behave properly in the field". On the other hand, he said: "If I place myself at the head of my Markish troops, even if I have lost half of my monarchy, as long as I still have my head, I'll drive the devil out of hell!"

On November 22, 1806 the regiment came to an end in Le Coq's Corps at Hamelin after the fall of Magdeburg, capitulating despite the protests of the officers and opposition of the men; the Third Battalion surrendered at Küstrin.

Infanterie-Regiment
Nr. 19
PRO GLORIA ET PATRIA
G. Dorn

INFANTRY REGIMENT NO. 20
Regiment on Foot

Commanders of the Regiment

1689		Major General Karl, Count von Schonberg
1698	7/3	Colonel Johann Heinrich von Börstel, later Lieutenant General
1711	1/30	Major General Ulrich Christoph von Stillen, later Lieutenant General
1728	12/10	Colonel Wilhelm Chenu de Chalezac de L'Aujardière
1732	1/3	Colonel David Jürgen von Grevenitz, later of I.R. 40
1741	8/31	Colonel Johann August von Voigt, later Major General
1742	8/18	Colonel Johann Kaspar von Hertzberg, died 1745 as Major General at Kesselsdorf
1746	1/10	Major General Franz Andreas von Borcke, later Lieutenant General
1756	7/20	Major General Berend Asmus von Zastrow, died 1757 at Aussig
1757	5/12	Major General August Gottlieb von Bornstedt
1759	4/7	Major General Otto Ludwig von Stutterheim, later Lieutenant General
1778	4/2	Major General Ludwig Karl von Kalckstein
1784	6/13	Major General Matthias Wilhelm von Below, later Lieutenant General and Governor of Stettin
1786	3/1	Major General Dietrich Eugen Philipp von Bornstedt, later Lieutenant General
1793	2/23	Major General Ludwig, Prince of Baden
1795	2/23	Major General Louis Ferdinand, Prince of Prussia, later Lieutenant General, died 1806 at Saalfeld

This regiment was established at Magdeburg in 1689, made of seven companies of Colonel de Hutten, two from Margrave Carl's regiment from Magdeburg and the Hutten Free Company. It was taken command of by Major General von Schonberg, son of Armand, Count of Schonberg, Marshal of France, and General-in-Chief of Brandenburg since April 27, 1687. It captured Kaiserswerth and Bonn. After giving troops to Duke Christian Albrecht von Holstein-Gottorp, the First Battalion went to Brussels in July of 1690, the Second remaining in Magdeburg and marching to Hungary in 1691 along with a company of the Varenne Regiment in 1691. On August 19 of that year it proved itself at Szlankamen: "By the quiet fire of the Hutten Battalion"—Colonel Bernhard de Hutten became its commander when Schonberg went into English service—"the wild attacks of the Spahis were broken", but in the battle Lieutenant Colonel Johann Friedrich von Kalckstein, its commander, fell. In the summer of 1692 it returned. In 1691 and 1692 the First Battalion was at Liege, Leuze, Namur and Luxembourg, then it fought before Huy and Liege, captured Namur in 1695 and stayed in Flanders until 1697. Four companies of the Second Battalion joined the regiment of Margrave Karl Philipp of Brandenburg for service in Piedmont. Only one company remained in garrison at Magdeburg; after the release of the field regiment in November of 1697 it was transferred to Colonel Johann Heinrich von Börstel on July 3, 1698. The two companies were enlarged to a battalion in 1706. On January 30, 1711 Major General von Stillen took command of the battalion, which received a Second Battalion on March 23, 1713, composed of three new companies and the Cornuaud (since 1686) and Vulson (since 1647) Free Companies. It was typical that cadres were enlarged again and again for political reasons. In 1715 one company of the Second Battalion landed on Rügen, and the regiment took Stralsund on December 23. In 1716 its garrison was Magdeburg. Its recruits came from Magdeburg and in part from Oebisfelde and Seehausen, as well as Forest Districts II and III, and part of I.

On December 16, 1740 it marched into Silesia and fought with its grenadiers at Mollwitz on April 10, 1741, in the first line at the end of the left wing, which made the victorious urn toward the enemy's left flank. Then Zander, a non-commissioned officer in disguise, reconnoitered the Fortress of Brieg, formerly besieged in vain, which fell quickly after bombardment on May 4 of that year, with 52 guns. Zahner was promoted to Captain. Under Schwerin on December 27 it helped to take Olmütz and remained there until the King arrived. Early in May of 1742 the whole regiment was assigned to secure Upper Silesia under the Old Dessauer. In the King's First Corps in 1744 it captured Prague on September 12. On February 14, 1745 it took the Plomnitz Heights east of Habelschwerdt under von Lehwaldt. In the battle of Hohenfriedeberg on June 4, 1745 it was on the left wing of the second line and, with the grenadiers, helped to carry the main attack on Günthersdorf. At the end of August it marched with Lieutenant General Count Gesslervia Reichenberg to the Old Dessauer's camp at Dieskau, and stormed the icy heights of Kesselsdorf on December 15 of that year. Here its chief, Major General Johann Kaspar von Hertzberg, fell in the right wing's first attack.

In the King's Army again in 1756 as an elite regiment, it took part in the capture of Tetschen Castle in mid-September before the First Battalion and grenadiers made the final attack on the city at Lobositz on October 1, "with high honors". After it marched into Bohemia again in 1757 and seized the city of Aussig, losing its chief, Major General von Zastrow, in a Croat attack, it took part in the siege of Prague and then shared in the defeat at Kolin on June 18 as part of the Pannewitz Brigade: At Manstein's command the Second Battalion attacked Croats at Chozenitz, pursued them too far in the direction of the Kamajka Heights, and had to be relieved by the First Battalion, an action completely counter to the King's plans. It held on to the end of the disastrous battle, losing 800 men! While the grenadiers moved toward Rossbach with the King, the regiment went to Silesia with the Duke of Bevern, experiencing the defeat at Breslau on November 22, so that only one battalion remained to fight at Leuthen on December 5. In the summer of 1758 it was among the troops surrounding Olmütz under the King's command. While the grenadiers took part in the advance toward Franconia, the regiment suffered the catastrophe of Hochkirch on October 14; the King sent it into a closed counterattack west of the village, where the Austrians surrounded it and drove it back. It lost 500 men. With the King's army at Landeshut in 1759, it spent the summer in camp at Schmottseiffen; the grenadiers fought well in the defeat of Kunersdorf and the victorious conflict at Korbitz on September 21, fighting with great bravery. It was always able to replace all losses from the Magdeburg area, being that district's primary troop. In Prince Heinrich's Corps from 1760 on—only the grenadiers fought at Liegnitz—the entire regiment (in the Ramin Brigade) fought at Torgau on November 3, advancing on the Süptitz Heights in the second attack of the first encounter, being driven back and standing fast under fire. It lost 600 men. In 1761 it was on the Mulde; in 1762, under its chief, Major General von Stutterheim, it took the battlements in the northern part of the Spittelwald; an abatis was stormed by 300 volunteers under Captain Georg Dietrich von Pfuhl, who received the Pour-le-merite for it.

The regiment's chiefs were changed often—three died in battle, others were removed "ungraciously": von Stutterheim and von Kalckstein—and it gained little prominence, even though Prince Louis Ferdinand, uncle of Friedrich Wilhelm III, gave it his name. It was disbanded in the Magdeburg capitulation on November 28, 1806.

Infanterie - Regiment
Nr. 20
PRO GLORIA ET PATRIA
G. Dorn

INFANTRY REGIMENT NO. 21
Regiment on foot

Commanders of the Regiment

1713		Major General Ernst, Count von Dönhoff, later Lieutenant General
1724	7/24	Colonel Heinrich Carl von der Marwitz, later General of the Infantry
1744	12/31	Major General Asmus Ehrenreich von Bredow
1756	2/25	Major General Johann Dietrich von Hülsen, later Lieutenant General and Governor of Berlin
1767	6/7	Colonel Carl Magnus von Schwerin
1773	1/11	General of the Infantry Carl Wilhelm Ferdinand, Crown Prince of Braunschweig, later Field Marshal

Per order of April 3, 1713, Major General Count Ernst von Dönhoff was given two companies each of Regiments 8 through 10, 13 and 17, just returned from the service of The Netherlands in the War of the Spanish Succession, to form a regiment to be put under Prussian administration. The regiment took part in the 1715 campaign, in the landing on Ruögen with its Second Battalion, and in the capture of Stralsund on December 23. In1725 it had one battalion each garrisoned in Halberstadt and Quedlinburg. When the "new garrisons" were established in 1729 as a kind of reserve system for occupation in case of war, it provided 400 fully trained men to Magdeburg. On May 1, 1735—like all regiments—it received a grenadier company for each battalion. In 1729 it was mobilized against Hannover. In 1734-1735 it was stationed on the lower Rhine in the course of the War of the Polish Succession. From 1722 to 1796 its garrisons remained Halberstadt for the staff, First Battalion and grenadiers, and Quedlinburg for the Second Battalion. Replacements for outstanding personnel came from the eastern Harz, the Principality of Halberstadt, County of Wernigerode, Diocese of Quedlinburg and Lordship of Derenburg as well as the cities in them, including Bleicherode, Sachsa, Ellrich, Beneckenstein, Hohenstein and Ilsenburg.

In 1740 the grenadiers, along with those of I.R. 5, at first under the command of Major von Götze, as of February 10, 1741 of Major Hans Carl von Winterfeldt, joined the Second Corps under the Duke of Holstein-Beck in the siege of Glogau, which they stormed on March 9, 1741. They were at Mollwitz on April 10 of that year, while the regiment was with Prince Leopold of Anhalt-Dessau's Observation Corps in camp at Göttin, south of Brandenburg/Havel, from early April to mid-September of 1741. In 1742 it was not on field duty. In the Second Silesian War it belonged to the First Army Corps of the King, which moved across Saxony into Bohemia from the northwest, in order to help encircle Prague early in September of 1744. When an Austrian relief army appeared from the Upper Palatinate, Major General Count Hacke was to take a fighting force, including the Second Battalion, and remove the magazine at Beraun on September 6. Attacks by the enemy's superior cavalry forces prevented this, to be sure, but could not break the "countenance and steadfast defense" of the three infantry squares. Prague surrendered on September 16. After the losses in the fall of that year, the King still had to suffer more losses while retreating over the mountains to Silesia. The decisive move was to take place early in summer: as of May 22, 1745 it took part in Margrave Karl's successful—with Zieten's help—march back from Jägerndorf to Neustadt O/S, and on June 4 it was with the Hertzberg Brigade on the left flank in the second encounter at Hohenfriedeberg, leading the main attack on Gunthersdorf. The grenadiers on the left flank had already taken Thomaswaldau under Major General von Polentz. They also fought bravely at Soor on September 30, but the regiment had already been marching to Saxony since the end of August to join the Old Dessauer's corps, which moved slowly toward Torgau early in December. In the Battle of Kesselsdorf, under Prince Moritz of Anhalt, it stormed the heights on the far side of the Kessel-Grunde at the end.

In 1756 it belonged to the King's army that went into battle at Lobositz on October 1; in the decisive final attack on Lobositz it lost twelve officers and 265 men. Colonel von Münchow, as regimental commander, and the two battalion commanders received the pour-le-merite at first for the "shining start". On October 12 Major General von Hülsen begged to speak "a most subordinate word for some of the captains, as I have found nothing to criticize in their accuracy in service and the duty and loyalty they showed at every opportunity". The King gave the regiment four more Pour-le-merite. Meanwhile the grenadiers were serving at Dresden and Aussig. After taking part in the encircling of Prague in 1757, the regiment marched with the King to the Battle of Kolin on June 18, where its commander, Major General von Hülsen, proved himself to be a brave troop leader of rank. Just at its high point it was attacked and smashed by an attack of Saxon and Austrian cavalry, losing eleven dead and sixteen captured officers, 500 dead, 200 wounded and 250 captured men! Lieutenant General von Treskow, commanding the left wing, was captured along with the colors of the regiment, which was all but wiped out. Only the First Battalion was present in the second encounter at Rossbach on November 5. At Moys on September 7 the grenadiers had lost 376 men. They covered the left flank at Leuthen. In 1758 the regiment joined Prince Heinrich's Saxon Corps and did a fine job of securing Saxony.

The grenadiers captured the fortress of Schweidnitz in a night attack on April 16, and led the right wing's successful counter-attacks at Hochkirch. At Greiffenberg on March 26, 1759 they were taken prisoner after fighting bravely against overwhelming odds. At Sebastiansberg in April of that year the regiment carried out a successful undertaking. In July it marched to the Oder, fighting at Kunersdorf on August 12, where it lost 25 officers and 783 men in the Kuhgrund and was nearly wiped out. What remained of the battalion was taken prisoner on November 21, still numbering fifteen officers and 582 men. In the winter of 1759-1760 it was able to reform with recovered invalids and recruits, and in the following spring it was already serving in Prince Heinrich's Corps, which went to Silesia that summer to march toward Torgau with the King in mid-October. The First Battalion fought in the Tettenborn Brigade of Zieten's Corps, which took Süptitz, while General von Hlsen, its chief, became "the hero of the day" at the last minute. It remained in the Saxon Corps in 1761, as also in 1762, as a "reliable support", without fighting at Freiberg.

When General of the Infantry Carl Wilhelm Ferdinand, Hereditary Prince of Braunschweig-Wolfenbüttel, Duke since 1780 and nephew of the King, became chief of the regiment in 1773, it took on a new luster, aided by its successes in Holland, as a result of which the Brandenburg Gate was erected in Berlin in 1787. Mortally wounded at Auerstedt at the age of seventy-one, he took the responsibility for the defeat of Prussia. The regiment was disbanded in the Prenzlau capitulation of October 28, 1806; the Third Battalion surrendered at Magdeburg.

Infanterie-Regiment
Nr. 21

INFANTRY REGIMENT NO. 22
Regiment on Foot

Commanders of the Regiment

1713	3/23	Major General Adrian Bernd von Borcke, later Count and Field Marshal
1741	5/31	Colonel Moritz, Prince of Anhalt-Dessau, later Field Marshal
1760	4/22	Major General Balthasar Rudolf von Schenckendorff, later Lieutenant General
1768	4/5	Major General Carl Christoph von Ploetz
1777	1/11	Major General Friedrich Carl von Schlieben, later Lieutenant General
1791	5/20	Major General Carl Friedrich von Klinckowström
1795	12/26	Lieutenant General Franz Otto von Pirch, later General of the Infantry

As ordered on March 23, 1713, the Third Battalion of the former Crown Prince Regiment No. 6, under its commander, Major General Adrian Bernd von Borcke, and French Free Company of Major General von Dorthe, the Marwitz garrison company in Oderberg (established in the Alt-Burgsdorff Regiment in 1626), and three newly recruited companies formed the regiment in Brandenburg/Havel; on July 1 of that year it was funded and also transferred to Pomerania. Friedrich II made Borck a count in 1740. Since August 25, 1714 it was stationed in Stettin, and as of April 10, 1715 it was already participating in the Pomeranian campaign: one company landed on Rügen with the First Battalion, the regiment helped to take Stralsund at year's end. In 1725 it was stationed in the Duchy of Pomerania, in 1729 with the corps mobilized because of Hannover, though its planned transfer to the lower Rhine in 1733-1734 did not take place. As the saying of the time, "The Prussians don't shoot that fast", shows, Friedrich Wilhelm I was very cautious with his preparations for war, but often felt himself disappointed and betrayed by the Emperor. In 1741 Prince Moritz of Anhalt-Dessau, youngest son of the "Old Dessauer", made it a prominent "Prince's Regiment". At Leuthen Prince Moritz commanded the first encounter, which had the job of leading the heavy flank advance. The King had good reasons for naming him Field Marshal on the battlefield: "You helped me more than anyone else has ever helped me before!" He had already proved himself well at Hohenfriedeberg, Kesselsdorf, Prague, Kolin and Berlin, and had also achieved prominence as a writer, thinker and colonizer of Pomerania. Badly wounded at Hochkirch and invalided out of the army, he received thanks from the King: "I only regret that I was not able to let you know my gratitude." Eighteen months later he died. The regiment stayed in Stargard from 1746 to 1806. Its recruits came from the populous, well-organized Principality of Cammin and the districts of Stargard, Neustettin, Belgard, Polzin, the cathedral town of Kolberg and the cities of Pyritz, Stargard and Bärwalde.

In 1740 it joined the King's First Corps, assembled until December 15 at the Crossen camp to march into Silesia and to Breslau. At Mollwitz on April 10, 1741 the First Battalion fought on the left wing in the first line, holding back an enemy attack under flank fire, then turning to achieve victory; the grenadiers fought there too after serving at Ohlau and Brieg. In 1742 it was with the King at Olmütz; from early May on it was in Upper Silesia with the Observation corps, under the Dessauer. In the King's First Corps again, it helped to take Prague and experienced the terrible march back to Silesia. At Hohenfriedeberg on April 10, 1745 it was on the right wing in the first line, under Lieutenant General Count Waldburg, carrying the charge toward the sector boundary of the united opponents, making the famous turn to the left under artillery fire, "very well executed", leading the bayonet attack over swamps and ditches and then making a frontal attack on Günthersdorf. Then it marched to Saxony to join the "Old Dessauer", who met the enemy at Halle in the autumn and at Meissen on December 12. Three days later at Kesselsdorf, on the left wing with the grenadiers, it stormed over the Kessel-Grund at the command of its chief. From 1746 to 1756 it was stationed in Stargard for training beyond its war experiences.

In 1756 it belonged to the army of the King; the regiment surrounded Pirna, the grenadiers fought at Lobositz. The siege of Prague spared it the battle, but it was hit hard at Kolin on June 18, 1757: when the army stopped because of the fighting at Krzeczhorz, Prince Moritz turned aside on his own initiative to take the Krzeczhorz Heights before the enemy had regrouped. Late in the afternoon the Austrian right wing had been penetrated, but the command for a unified advance with cavalry and reserves did not reach the Duke of Bevern. His regiments, including I.R. 22, were already being drawn into defeat under heavy artillery fire between Brzezan and Chozenitz. When the enemy cavalry then attacked, it retreated with its few hundred men. Twenty-six officers and 1165 men lay on the battlefield, four-fifths of its manpower! The grenadiers fought at Rossbach and Leuthen. Those taken prisoner at Kolin returned at the beginning of 1758, when the regiment, at full strength again, was in the Pomeranian Corps that joined the King on August 22 and met the Russians three days later at Zorndorf, along with the grenadiers. On the right wing under Count Dohna, it met the Russian cavalry's counterattack in the midst of its own attack, but the victory was won at a high cost. The regiment then served in less important actions, such as the relief of Kolberg at the end of October. Only the grenadiers went into battle at Kay on July 23, 1759; the regiment took Frankfurt/Oder under Wobersnow. While the grenadiers shed their blood at Kunersdorf, the regiment, under Kleist, marched from Pomerania to join the King at Fürstenwalde. With the King again in 1760, it unsuccessfully besieged Dresden from July 10 to 22. Meanwhile the grenadiers had fought to the last at Landeshut. At Torgau on November 3 it was originally in reserve behind the cavalry. As night came on, the First Battalion, trying to bring back the wounded, met the last counterattack and gave Zieten a hand, while the Second remained in reserve. Cannons, flags and rich booty fell into his hands. Both commanders received the Pour-le-merite on November 12. With the King again in 1761-1762, it came from Ludwigsdorf under Count Lottum and stormed the northern battlements, won the left flank and captured the works. Four company commanders received the Pour-le-merite. It had regained its status again.

It was part of the main army in 1806. The First Battalion was taken prisoner at Nieder-Rossla on October 15, the Second in the capitulation of Pasewalk on October 29, the Third in Stettin. The grenadiers gathered in Kolberg and formed the Waldenfels Battalion, later the First Emperor Alexander Guard Grenadier Regiment No. 1. The canton troops formed the First Pomeranian Regiment.

Infanterie-Regiment
Nr. 22

INFANTRY REGIMENT NO. 23
Regiment on Foot

Commanders of the Regiment

1713	May	Major General Paul Anton von Kamecke
1716	Feb.	Colonel Johann Quirin von Forcade de Biaix, later Lieutenant General and Commandant of Berlin
1729	3/27	Colonel Egidius Ehrenreich von Sydow
1743	7/24	Major General Alexander Ernst von Blanckensee, died 9/30/1745 at Soor
1745	10/30	Major General Christoph, Count zu Dohna, transferred to I.R. 16
1748	7/14	Major General Friedrich Wilhelm von Forcade de Biaix, later Lieutenant General
1765	5/14	Colonel Christian Ernst von Puttkamer, transferred to Rentzel's Garrison Regiment
1766	5/31	Major General Christoph Friedrich von Rentzel, later Lieutenant General
1778	11/7	Colonel August Wilhelm von Thüna, later Major General
1786	3/1	Major General Stephan von Lichnowski, later Lieutenant General
1796	12/6	Colonel Christian Ludwig von Winning, later Lieutenant General

This regiment too was established upon the accession of King Friedrich Wilhelm I, who regarded himself as "his own finance minister and field marshal", formed of six companies of Grenadier Guards (No,18), and four of Fusilier Guards (No. 1) in May of 1713 and put under the command of Major General Paul Anton von Kamecke on June 1. In the Pomeranian campaign of 1715 one company of the Second Battalion took part in the landing on Rügen as well as the capture of the Peenemünde battlements, and the regiment fought at Stralsund. In February of 1716 Colonel Johann Quirin von Forcade, a Huguenot, took command of the regiment, which was stationed in Königsvorstadt and Spandau, in the suburbs of Berlin, from 1716 to 1806. He was its chief for thirteen years, and his famous son led it for seventeen years. The military experience which they brought with them from the French army, plus reformed concepts of duty and Markish replacements from the districts of Niederbarnim, Stolpe-Uckermark and in part Teltow, with the cities of Oranienburg and Liebenwalde, made it into a group of "elite origins". Frederick the Great said: "If I want to see soldiers, I must see this regiment!" Friedrich Wilhelm I promoted discipline without any differentiation, likewise righteousness and proper treatment of the soldiers, his "dear blue children".

After gathering at the Crossen camp until December 15, it marched to Breslau with the King, on the right wing of the First Corps. On January 25, 1741 it saw action at Gratz, south of Troppau, then on April 10 it fought on the left wing of the second line in the Battle of Mollwitz, losing 93 men, and from October 18 to 31 of that year it took part in the second siege of the Fortress of Neisse, which capitulated according to the "Secret Treaty of Schnellendorf". In 1742 it went with the King to Moravia, saw service at Olmütz, but stayed in Kamenz and Wartha till the end of the campaign to secure the County of Glatz. In the fall of 1744 it went to Bohemia with the Second Corps of the King and survived the terrible winter retreat to Silesia. In the spring of 1745, under Major General von Hautcharmoy, it cleaned Hungarian raiders out of Upper Silesia between Ujest, Gross Strehlitz, Konstadt and Kreuzburg, encouraged by the King: "Whoever is not bold and brave does not deserve to serve in the Prussian Army". At Hohenfriedeberg on June 14 of that year it took part in the big attack on Günthersdorf in the first line, fighting next to the Guards on the inner left wing, suffering light losses. But at Soor on September 30 it lost its commander, Major General von Blanckensee, his son and adjutant, in the first storming of the Graner Koppe, which the regiment then took in an attack of the second encounter, against the enemy's counterthrust, this time with heavy losses. The regimental commander, Colonel Friedrich Wilhelm Quirin von Forcade, received the Pour-le-merite on January 6, 1746, as did eight company commanders at the "Special Review" at the Zoological Gardens on May 27, 1747, almost two years later, "in uncommon gracious, generous and moving expression of all-highest satisfaction, as well as true esteem, in that His Majesty distributed the Order of Pour-le-merite to all captains of the battalions at Soor".

In the summer of 1748 Forcade, already a commander since the early Forties, became its chief, In 1744 he had denied Ensign von Wulffen satisfaction in the matter of a dispute, and the ensign later forced his way into Forcade's quarters at sword's point and injured him. The court-martial passed a death sentence. The King decided that a subaltern could only demand satisfaction afterhis service from then on, a principle which has remained to this day. After proving himself at Lobositz and Pirna in 1756, he was severely wounded at Zorndorf. In 1760 he led a corps in Far Pomerania and commanded the right wing of the second line at Torgau; he had a leading role until 1762.

In 1756 the regiment joined the Saxons in the siege of Pirna; on April 21, 1757 it penetrated Bohemia at Reichenberg, and on May 6 it was on the right wing of the first line when it broke through south of Kej at Prague, leading to the bloody infantry battle on the Rokenitz-Grund. It lost 22 officers and 602 men. Afterward it took part in the siege of Prague; then came the march to Rossbach, where on November 5 only about one-third of the infantry got into the battle. In a quick march it returned to Leuthen, where the victory was due above all to the cooperation of the lower leadership. In 1758 the grenadiers stood out at the storming of Schweidnitz on April 16. Afterward it marched with the King to Olmütz. On account of the Russians, he turned northward to the Oder as of August 11. At Zorndorf on August 25 it and other regiments overcame the serious crisis in the afternoon and swung to the left, driving the enemy back. During the last attack, about 7:00 P.M., General von Forcade was carried away, badly wounded. The regiment lost 400 men and had only 1000 left. At Hochkirch seven weeks later its counterattack was at first successful. When Prince Moritz led it again, he was severely wounded. Archenholtz reports: "The Forcade Regiment was one of these groups that knew of lost battles only from hearsay. Many had such a high sense of honor that they did not want to give in to superior forces. Others had to be dragged off the field by force because they did not want to survive a day of misfortune." Now it had only 600 men left. In 1759 it was in camp at Schmottseiffen; it was spared Kunersdorf. In 1760 the regiment and grenadiers fought on the left wing at Liegnitz, winning two Pour-le-merite, and the grenadiers fought at Hochgiersdorf on September 17. At Torgau on November 3 it fought on the left in the first line, losing its commander, Colonel von Butzke, 22 officers and 365 men in the third attack. In 1761-1762 it remained in Silesia.

Beaten at Jena in 1806 with Rüchel's Corps, it was disbanded in the Erfurt and Prenzlau capitulations as well as being defeated in battle at Lübeck in Blücher's Corps, while the Third Battalion surrendered in Stettin. The canton troops joined Kolberg's Regiment.

Infanterie-Regiment
Nr. 23
PRO GLORIA ET PATRIA
G. Dorn

INFANTRY REGIMENT NO. 24
Regiment on Foot

Commanders of the Regiment

1715	12/25	Major General Johann Sigismund, Baron von Schwendy, later Lieutenant General
1723	1/13	Major General Curt Christoph von Schwerin, later Count and Field Marshal, died 5/6/1757 at Prague
1757	5/11	Major General Carl Christoph, Baron von der Goltz, later Lieutenant General, died 6/30/1761
1761-	1763	vacant
1763	4/8	Colonel Bernhard Alexander von Diringshofen
1776	1/12	Colonel Leopold, Prince of Braunschweig, drownedin a rescue attempt in the Oder on 4/27/1785 as Duke and Major General
1785	5/26	Major General Gottlieb Ludwig von Beville, later Lieutenant General
1791	5/20	Major General Carl Ludwig Sylvius von Franckenberg
1795	7/29	Colonel Johann Ludwig von Grünberg, later Major General
1799	2/5	Colonel August Wilhelm von Zenge, later Major General

This regiment came into being in the course of new organization in 1713-1715, made of field battalions that had been created of garrison troops on June 1, 1713 for Major General Johann Sigismund, Baron von Schwendy, comprising two Spandau companies (since 1637), one Frankfurt company (since 1655), and one company each of Fusilier Guards and Engineers, later the First Battalion, and for Major General Berend Christian von Schnebeck in Kolberg (since 1648), later the Second Battalion. Two days after the taking of Stralsund in 1715, the two battalions joined as one regiment under von Schwendy. In Major General Curt Christoph von Schwerin it had a commander who gave it its character. He was a man of the world, a cavalier, a builder, the opposite of the "Old Dessauer" and in the King's judgment a "very brave and capable, but too lively man", but after all, the victor at Mollwitz. In 1729 the regiment contributed 400 men as a "new garrison" for Stettin; in the same year it was mobilized against Hannover. At this time the colors and symbols of all regiments were recorded in writing for the first time. Its garrison was always Frankfurt/Oder, with the grenadiers at Cottbus until 1756 and Fürstenwalde afterward. The recruits came from the districts of Züllichau, Crossenand in part Cottbus with the cities in them, plus Rotenburg and Bobersberg.

On the march into Silesia on December 16, 1740 it belonged to the right wing of the First Corps. Under its commander, Lieutenant Colonel von Tettau, it marched to Neisse. At Mollwitz on April 10, 1741 he urged the King to let him conduct the battle because of the superiority of the Austrian cavalry—a unique approach! On the left wing of the first line, it enjoyed a victory after the turning of the fresh wing, with only forty losses. On August 10 Schwerin had to have Breslau occupied because of anti-Prussian activities. The King ordered him to prevent any oppression of the people. Then it left Upper Silesia with Schwerin's Corps, bound for northern Moravia, where it took Olmütz on December 27. In 1742 the First Battalion and the grenadiers secured Upper Silesia and Moravia, the Second stayed at Olmütz, until it went to northern Bohemia with the King at the beginning of May. Under massive attack at Chotusitz on May 17, 1741, it stood firm during fighting in numerous places, between burning buildings; according to the Duke of Bevern, the "great inner orderliness introduced by Friedrich Wilhelm was markedly superior to the experience of the Austrian generals". Peacetime brought ongoing engineering training and, as of the fall of 1743, even mixed weapon drill. In 1744 it besieged Prague in the First Corps from September 11 to 16. The grenadiers stormed the Ziska battlements; the King promoted Grenadier Krauel to Lieutenant because he was the first man in. There followed the awful march back through snow, rain, storm and coldness, plus "lack of sustenance in the land". In the spring of 1745 helped to clean out Upper Silesia. After a rest, it attacked with the grenadiers and next to the Guards in the first line at Hohenfriedeberg on June 4, first at the Gule, then at Günthersdorf. On November 23 it fought at Kath. Hennersdorf.

In 1756 it marched on the right wing, under the Duke of Bevern, through Reichenberg to Pirna and took part in the siege. At Prague on May 6 it was on the left wing, leading the first attack on the Homole-Berg, under Winterfeldt, who was soon wounded, fighting under heavy rifle and cannon fire until it came close to the enemy, only to be driven back. Disorder resulted. The 73-year-old Schwerin charged ahead of the cavalry and tore a flag from the hand of Staff Captain von Rohr to lead the regiment forward again. Then five balls of grapeshot hit him fatally. The flag was his shroud. Lord Mitchell, the English emissary, praised him as "one of the greatest officers that this land has produced, and one of the best men". He had not wanted the battle. The regiment lost thirteen officers and 522 men. After breaking through the center, the King led Schwerin's wing to Prague to take part in the siege. The regiment fought at Rossbach, the grenadiers at Leuthen. In 1758 it was with the Saxon Corps with Prince Heinrich, capably securing the southwest flank in moving, continual fighting. In 1759 it moved into Franconia with him, but soon was sent under Hülsen to strengthen the Pomeranian Corps, losing 37 officers and 933 men in the defeat at Kay, one-sixth of the total losses. At Kunersdorf three weeks later it took part in the infantry battle on the Kuh-Grund. In 1760 the King led it to Dresden, the siege of which he had to break off after twelve days on July 22. Marching without rest from the Elbe to Bunzlau in five days, it took part in the victory of Liegnitz on August 15, preventing the Russians and Austrians from joining forces, and on September 17 the Battle of Hoch-Giersdorf, while the grenadiers fought bitterly to the last man under Fouque at Landeshut on June 23. The battalion was wiped out. At Torgau on November 3 the regiment, on the first line, took part in the second attack on the Süptitz Heights; it succeeded but was then stopped by the enemy reserves, with the loss of ten officers and 699 soldiers. In 1761 the regiment was stationed on the Mulde in Saxony, opposite the Austrians and the Imperial Army. Its commander, Lieutenant General Baron von der Goltz, commander of a corps that was planning to take the main Russian magazine in Posen, died suddenly on June 30, 1762. On July 20 the grenadiers, along with the Guards, captured Ohmsdorf Castle as the prelude to the taking of the Burkersdorf battlements. At Freiberg on October 29 the regiment, in Colonel von Diringshofen's Brigade, stormed the heights of St. Michael northwest of Brand.

With the main army in 1806, it was beaten at Auerstedt and disbanded in the capitulations of Erfurt and Pasewalk on October 29, while the Third Battalion surrendered in Küstrin. The men on canton duty joined the First West Prussian Regiment.

Infanterie-Regiment
Nr. 24

INFANTRY REGIMENT NO. 25
Regiment on Foot

Commanders of the Regiment

1715	9/27	General of the Infantry Otto, Baron von Schlabrendorff
1718	Aug.	Colonel Johann Christoph, Count von Wylich and Lottum, later Major General
1727	10/22	Colonel Martin von Thile
1728	2/18	Major General Friedrich August, Count von Rutowski
1729	3/29	Colonel Christoph Wilhelm von Kalckstein, later Field Marshal, died 6/2/1759
1760	2/8	Major General Friedrich Ehrentreich von Ramin, later Lieutenant General and Governor of Berlin
1782	12/13	Lieutenant General Wichard Joachim Heinrich von Möllendorff, later Field Marshal, Vice-President of the Upper War College, Governor of Berlin

The field battalion of Lieutenant General Otto, Baron von Schlabrendorff, established on June 1, 1713 and composed of the Küstrin garrison and the Driesen Free Company, both in existence since 1626, and the field battalion of Major General Anton von Pannewitz, composed of the two garrison companies of Peitz (north of Cottbus), the two of Memel, all dating back to 1626, and the garrison company of Friedrichsburg (since 1670), were by order of September 27, 1715 united to be the First and Second Battalions of a regiment commanded by General of the Infantry von Schlabrendorff. It came into being shortly before I.R. 24. One company landed in the Third Battalion on Rügen. On October 20, 1715 the regiment opened the communication lines to Stralsund. Under the Count of Wylich and Lottum since 1718, it was stationed in Pomerania in 1725. In 1728 it received Major General Count von Rutkowski as its chief, an illegitimate son of King August the Strong and the Countess of Cosel. He returned to Saxony after sufficient experience with the training of the Prussian army, and in 1756, as Field Marshal, he became the commander of the Saxon troops encamped at Pirna. He was succeeded in 1729 by the 47-year-old Colonel von Kalckstein, since 1718 one of the teachers of Friedrich II, who commanded the regiment for thirty years and was a great soldier, calm and self-controlled, an exemplary instructor for the regiment. He became Field Marshal in 1747, as did in 1793 his successor von Möllendorff, who had risen from Major to General in three years. Since 1729 the regiment was stationed in Friedrichstadt, a southern suburb of Berlin. The districts of Lebus and Sternberg/Neumark, with those two cities, provided the recruits, as did, from 1773 on, parts of the West Prussian border region of Deutsch-Krone and Hohensalza as far as Bromberg. From 1729 to 1806 each chief led this regiment for twenty-two years or more.

It marched to Silesia only in February and March of 1741, in the Third Corps. The grenadiers had participated in the storming of Glogau on March 9; the regiment went into battle at Mollwitz on April 10. Here the First Battalion fought on the inner right wing of the first line, and was thus not much harmed by the enemy cavalry attack. The grenadiers held fast against a storming to secure the right flank between the two lines, preventing a catastrophe. The First Battalion lost two officers and 123 men. Major Wolf von Retzow received the Pour-le-merite. In October of that year it moved to northern Bohemia under Hereditary Prince Leopold. In 1742 it occupied Pardubitz and Königgrätz, holding them without going into the field. Pardubitz was an important supply base. In 1744 it was a vital part of the King's First Corps and suffered the early-winter withdrawal under constant attack by the enemy's pursuing light troops. At Hohenfriedeberg on June 4, 1745 it fought at the middle in the first line, in Blanckensee's Brigade' Kalckstein commanded the second line, which included the attack that succeeded after the Prussian cavalry attack, until the Bayreuth Dragoons made the last move. The enemy's moves in Bohemia were countered by the King at Soor on September 30. When the regiment passed by Burkersdorf, he ordered the Second Battalion to move in cautiously so as to prevent the enemy from settling there. From this there developed an unwanted fight on the left wing, which drew the regiment into an enemy attack and artillery fire. But it moved on unshaken, losing 264 men. At the review on May 26, 1747—Kalckstein was already Field Marshal-Colonel von Woedtke and four company commanders received the Pour-le-merite.

In 1756 the regiment was involved in the surrounding of the Saxons at Pirna—almost a reunion with Count von Rutkowski. The Saxons surrendered on October 16. The siege of Prague followed in 1757. Between the besieged Prague and the superior main force of the enemy on the heights west of Kolin, the King decided to attack on June 18. Bevern's Division, which did not receive the King's command to march to the left in time, defended itself again and again that afternoon, from Brzezan to Chozenitz, with the greatest bravery, until it had lost a third of its officers. In mid-August the King left Bohemia, but then sent the Duke of Bevern—with the regiment—to Silesia, where on November 22 it came under attack at Breslau. For hours 27,000 Prussians defended themselves against 83,000 Austrians, repeatedly fighting off their attacks, but the regiment lost its commander, Colonel Hieronimus Wilhelm von Eckardt. But with astonishing steadfastness it bore both losses and, on December 5 at Leuthen, fought its way around the village, with the grenadiers fighting bitterly. After that it besieged Breslau. In 1758 it joined the King and went to Olmütz, then via Königgrätz to Landeshut, and from there to the Oder on August 11, where on August 25 it stood fast at Zorndorf, holding off the massed Russian cavalry attack on the right flank and overcoming the stubborn resistance of the Russian infantry step by step. It also helped out at Hochkirch, protecting the retreat after the Retzow Corps had neglected to take the Strohm-Berg. In 1759 it was encamped at Schmottseiffen with the King, while the grenadiers were taken prisoner at Maxen at year's end after having suffered heavy losses in an attack on the Russian batteries at Kunersdorf. With Prince Heinrich's Corps in 1760, it took part in the Battle of Torgau on November 3, fighting in the Queiss Brigade on the left wing of the second line as the king tried to take the Elbe Line away from the enemy. It marched on in the second column under von Bülow and took theSüptitz Heights, broke through but was thrown back. In 1761 it was encamped at Bunzelwitz, then headed for Pomerania on October 25, stormed the Kolberg batlements at Spie, and was in Saxony again by year's end. Back in Silesia in 1762, it fought in the Battle of Burkersdorf on July 21; behind the Lottum Brigade in the right storming column, it captured the northern battlements at Ludwigsdorf. Afterward it besieged Schweidnitz.

In the main army in 1806, it was disbanded in the Prenzlau capitulation on October 28; the Third Battalion surrendered in Stettin. The men on canton duty went to the First West Prussian Regiment.

Infanterie-Regiment
Nr. 25
PRO GLORIA ET PATRIA

INFANTRY REGIMENT NO. 26
Regiment on Foot

Commanders of the Regiment

1714	12/10	Major General Curt Hildebrand, Baron vonn Löben, later Lieutenant General
1730	3/22	Colonel Henning Alexander von Kleist, later Field Marshal
1749	8/24	Major General Dietrich Richard von Meyerinck, formerly Commander of the Guard Regiment
1758	1/25	Major General Carl Heinrich von Wedell, later Lieutenant General
1760	12/25	Major General Christian Bogislav von Linden
1764	12/29	Colonel Anton Abraham von Steinkeller, later Major General and Commandant of Berlin
1778	3/25	Major General Hans Christoph von Woldeck, later Lieutenant General
1789	7/12	Major General Philipp Albrecht von Schwerin
1795	6/4*	ajor General Carl Leopold von Larisch, later Lieutenant General

As ordered on December 10, 1714, a regiment was established for Major General Curt Hildebrand, Baron von Löben as of January 1, 1715, composed of the field battalions of Baronvon Heiden as First and Count Waldburg as Second Battalion. The first had been formed in 1708, made of free companies and guard companies of the Alt-Holstein, Derfflinger, Zieten and Spaen Regiments that had been disbanded in 1697 after being in existence for twenty to forty years. It had fought in The Netherlands since 1710. The Second Battalion grew from the "Bodyguard of the Governor in Prussia" of Prince Radziwill, founded in 1659, commanded by the Duke of Croy as of 1670, expanded to a battalion in 1683, led by Count Carl Heinrich, Hereditary Steward of Waldburg, in 1705 and serving in Italy from 1708. Count zu Waldburg was President of the Königsberg War Commission in 1714, then Lord President of the Chamber and Leader of the "Retablissement" in East Prussia. In Stettin in 1715, this company landed on Rügen in the Fifth Battalion. In 1729 it provided 200 men as a "New Garrison" for Küstrin and was mobilized against Hannover. In 1733 its move to the lower Rhine in the War of the Polish Succession ever went beyond the planning stage. In 1730 Colonel von Kleist became the regimental chief, serving as such for nineteen years. In 1740, as a Major General, he surrounded Brieg; in 1747 the King made him, now Corps Commander, a Field Marshal. From 1716 on the regiment was garrisoned with six other regiments (not counting the Potsdam Guards) in Berlin. Its recruiting area comprised the districts of Lebus, Cottbus and Beeskow, with the cities of Müllrose, Seelow, Fürstenwalde and Peitz.

On December 16, 1740 the regiment marched to Silesia on the right wing of the First Corps under its commander, Lieutenant Colonel de Hautcharmoy. In the Battle of Mollwitz on April 10, 1741 it was on the inner right wing of the first line and was overrun by the Austrian cavalry; next to the two battalions on the right was the First Guard Battalion, whose platoon fire was "as calm and sure as on the drill field", so that the King could ascribe the subsequent saving of the victory by Schwerin to them. It lost 25 officers and 667 men. Its commander, a major, a captain and a lieutenant received the Pour-le-merite. In a report it was stated: It was almost impossible to restrain the man who, his bayonet fixed, wanted to attack the already retreating enemy with all his power", as happened in Schwerin's counterattack. In 1742 the regiment secured Upper Silesia and northern Moravia under the Old Dessauer, especially in Neisse and Neustadt, while the grenadiers were in Ottmachau. In 1744 it took part with the King's First Corps in the unfortunate Bohemian campaign, marching back to the Elbe in the fall. Most of the soldiers had dysentery and nothing to eat. Many deserted. In 1745 the regiment secured Silesian fortresses. On May 22 the grenadiers successfully defended the Landeshut Heights, and on June 4 they fought at the end of the right wing of the second line at Hohenfriedeberg. In 1749 the regiment received General von Meyerinck as its chief; as of 1740 he was still a captain in the First Guard Battalion. As a Lieutenant Colonel he had worked on the new Infantry Regulations in 1743.

In the King's Army in 1756, it was soon to prove its ability as one of the best regiments. At first it took part in surrounding the Saxon camp at Pirna. On May 6, 1757 it joined the battle at Prague. After the failure of the left wing's attack at Sterbohol under Winterfeldt, the King spotted a large gap in the center of the Austrian line, against which he sent the first line of the left wing under Hautcharmoy. He had the regiment break through south of Kej with heavy artillery support. So it came into the bitter fight at the Maleschitz Gorge. General de Hautcharmoy fell. After Kolin, which the regiment was spared, it went to Rossbach with the King on July 21, won a victory on November 5, then quick-marched to Breslau. At Leuthen on December 5, both of its battalions served as advance guard under Wedell, attacking the Sagschützer Kiefernberg. Prince Moritz of Dessau had been recommended for this task: "Your Majesty can entrust crown and scepter to this regiment; if they run from the enemy, I wouldn't want to be there either". At the beginning of the famous swing to the left, the King personally directed the regiment: "You must not advance too strongly, so the army can follow. You have the enemy before you and the whole army behind you!" In the attack on Leuthen Prince Moritz called out: "Honor enough, go back for the second line!" Our boys answered: "We would be dog food (if we did); give us bullets!" and beat back a cavalry attack, as Barsewisch reports. The regiment lost ten officers and 454 men, and on December 8 it received fourteen Pour-le-merite, very uncommon for one battle, exceeded only by Alt-Braunschweig at Rossbach. In 1758 Major General von Wedell took command of the regiment; he was a man of great experience and clear thinking, favored by the King as his "pupil": "What he is to do, he always carries out very well, better than I could hope". In 1761 he became Minister of War. Before Olmütz from June 1 to mid-July, the regiment was subsequently in Silesia, at Dresden on September 11, at Hochkirch on October 14; the King positioned it northwest of the village for a successful counterattack and held position behind the regiment for a long time, until cavalry cut it down to 150 men with three flags, who fought their way out. In 1759 it was with the King in camp at Schmottseiffen, while the grenadiers were in Franconia, at Kay and Kunersdorf. From July 10 to 22, 1760 it besieged Dresden, and on August 15 it fought at Liegnitz. At Torgau on November 3 it led the third attack in Butzke's Brigade, which shed blood until Zieten arrived. In 1761 it was in Saxony, opposing superior enemy forces. In 1762, in Diringshofen's Brigade, it stormed the heights of St. Michael. It was always in the King's Army until 1760.

With Rüchel's Corps in 1806, it was disbanded in the Erfurt capitulation on October 16; 9000 were liberated by Lieutenant Hellwig on October 17, and the majority was with Bücher's Corps at Ratekau on November 7; the Third Battalion surrendered in Stettin. The men on canton duty joined the First West Prussian Regiment.

Infanterie-Regiment
Nr. 26

INFANTRY REGIMENT NO. 27
Regiment on Foot

Commanders of the Regiment

1715	11/25	Colonel Leopold Maximilian, Prince of Anhalt-Dessau, later Field Marshal
1747	4/22	Major General Franz Ulrich von Kleist, later Lieutenant General, died 1/13/1757 of wounds at Lobositz
1757	1/20	Major General Moritz Wilhelm von Asseburg
1759	3/18	Major General Daniel Georg von Lindstedt
1764	7/27	Major General Peter Heinrich von Stojentin
1776	10/20	Major General Alexander Friedrich, Baron von Knobelsdorff, later Field Marshal
1800	10/2	Colonel Friedrich Wilhelm Alexander von Tschammer, later Major General

Even before the taking of Stralsund in the 1715 Pomeranian campaign, the King commanded on November 25 of that year that a new regiment be founded of Swedish prisoners from Rügen, some 400 German mercenaries. The majority of the officers, many junior officers and a body of privates came from the Guard Regiment No. 6 and I.R. 3 from Anhalt. Its commander was the second son of Prince Leopold of Anhalt-Dessau, Prince Leopold Maximilian, who became a Field Marshal in 1742, and it was considered a "Swedish-Pomeranian" regiment although the nationality scarcely mattered any more. It was in the Kurmark in 1725, and in 1729, the year of mobilization against Hannover, it provided 400 men for a "New Garrison" in Berlin. On May 1, 1735 it was given, as were all regiments, one grenadier company per battalion, which had been divided among the companies previously, without flags. The commander was allowed to seek replacements in the battalion, "only capable fellows". Under the "Soldier King" a problematic recruitment of foreigners began in 1721. Most of them were Germans who found garrison duty less than completely fulfilling. These were to be released by their own country, which was meant to attract tall men in particular. The result was desertion, as was punishment of criminals. On the other hand, the movement of officers from land to land was a common practice. The treatment of invalided or retired soldiers by the civil authorities, by order of the King, was exemplary. The regiment was garrisoned at Stendal, with the Second Battalion at Gardelegen; replacements came from the Altmark with the districts of Salzwedel, Arendsee, Tangermnde, Stendal and their cities, plus Bismarck, Apenburg and Beetzendorf. Prince Leopold was the regimental chief for 32 years.

Only in March of 1741 did it move into Silesia with the Third Corps, taking part on March 9 in the capture of Glogau, which had blocked the advance to the Oder. At Mollwitz on April 10 the grenadiers held their ground on the shattered right wing, though decimated to half their numbers. Simultaneously the regiment was on the inner left wing of the first line, which later decided the battle under Schwerin. In the process it lost seven officers and 309 men, mainly from the First Battalion. There followed months of training and evaluation of the experience gained at Mollwitz, not surprising after thirty years of peace. In October it went to northern Bohemia under its commander. The grenadiers were in Upper Silesia in 1742; the regiment went to Bohemia with the King at the end of April, followed by the grenadiers. At Chotusitz on May 17 it was attacked south of town by heavy forces with artillery support. Vigorous street fighting and cavalry attacks made the situation more difficult, to the extent that Chaplain Seegebart gathered those who retreated, spurred them on and led them forward again, until Wedell finally brought help. Five Pour-le-merite, including one to Colonel Ernst-Ludwig von Goetze as Regimental Commander, expressed the King's thanks. Fourteen officers and 428 men had fallen.

When war broke out in mid-August of 1744, the regiment stayed in Anhalt at first, then saw service in the Old Dessauer's Corps against the Saxons late in April of 1745. At Hohenfriedeberg the Grenadiers took Thomaswaldau but lost their commander, Colonel von Kahlbutz. To the west the Old Dessauer moved back to Meissen via Halle to strike Kesselsdorf on December 15. The grenadiers opened the attack on the right wing, losing Major von Kleist, their commander. The regiment fought for the heights in close combat, losing twelve officers and 451 men, including its commander, Colonel Friedrich Christoph von Rintorff. On May 27, 1747 it was awarded four Pour-le-merite.

With the King's Army at Lobositz on October 1, 1756 it was at the end of the left wing on the Lobosch-Berg, which absolutely had to be held. After five hours under fire, non-commissioned officers and men broke into an attack spontaneously and stormed to the shore of the Elbe: "Our Brandenburgers smashed the Pandurs like furies", it was said. Lieutenant General von Kleist, its chief, stayed on his horse despite the wounds that cost him his life three months later.; thirteen officers and 277 men fell. He received the Order of the Black Eagle, as did his majors, and seven captains were given the Pour-le-merite on January 25. Two weeks later the Saxons surrendered at Pirna. In 1757 it took part in the siege of Prague, but was spared Kolin. After the retreat to Leitmeritz it marched, under the Duke of Bevern, through Saxony to Silesia, where it held a fortified position on the east bank of the Lohe before Breslau for almost two months, beginning on September 26. The grenadiers fought at Moys on September 7. On November 22 the Austrians broke through the very bravely defended but poorly manned line, putting the possession of Silesia at risk. The regiment lost thirteen officers and 687 men. Then the King appeared on December 2. With the barely 500 men who remained of the battalion, it formed the avant-garde on the right wing at Leuthen on December 5, under Prince Carl of Bevern. The Commander, Lieutenant Colonel Friedrich Wilhelm von Wietstruck, fell. In 1758, in spite of everything, the King took it along to besiege Olmtz; in the summer it moved from there past Königgrätz, and on August 11 from Landeshut to the Oder, where it withstood the Russian mass attacks on the right wing at Zorndorf on August 11, and finally, with its last strength, was able to swing to the left against the enemy's last feeble opposition. This cost it twelve officers and 623 men, including its commander, Lieutenant Colonel Carl Ludwig von Zessen. Then it moved via Küstrin and Dresden to Silesia, while the grenadiers shed the first blood on the right wing at Hochkirch on October 14. At Greiffenburg on March 26 they were captured by 8000 enemy forces. The regiment was encamped at Schmottseiffen. With Prince Heinrich's Corps in 1760, then again with the King, it did not take part in any major action. In 1763 it had 1738 Prussians, 67 Saxons and 165 foreigners in its ranks.

The next-to-last regimental chief was highly esteemed by the King and described as one of the "most impressive figures of the Eighteenth Century". In 1806 the regiment was with Rüchel's Corps, then with Blücher's. It was disbanded in the fighting at Lübeck on November 5 and the capitulation of Ratekau on November 7, the Third Battalion at Hameln. Garrison troops joined the Bodyguard Regiment.

Infanterie-Regiment
Nr. 27
PRO GLORIA ET PATRIA
G. Dann

INFANTRY REGIMENT NO. 28
Regiment on Foot

Commanders of the Regiment

1723	1/3	Major General Konrad Wilhelm von der Mosel, later Lieutenant General
1733	9/14	Colonel Friedrich Ludwig, Count zu Dohna, transferred to I.R. 44
1742	1/30	Colonel Heinrich Carl Ludwig Herault de Hautcharmoy, later Lieutenant General, died 5/16/1757 of wounds at Prague
1758	1/5	Major General Lorenz Ernst von Münchow, diedJan. 1758 of wounds at Leuthen
1758	4/4	Major General Johann Friedrich von Kreytzen
1759	5/20	Major General Friedrich Ehrentreich von Ramin, transferred to I.R. 25
1760	2/8	Major General Friedrich Wilhelm von Thile
1770	6/4	Major General Michael Konstantin von Kalinowa Zaremba, later Lieutenant General
1786	9/6	Lieutenant General Ludwig Carl von Kalckstein, later Governor of Magdeburg, transferred to I.R. 5
1789	8/28	Colonel August von Grenzkow, later Major General
1792	12/29	Major General Friedrich Leopold von Rüts, transferred to I.R. 8, later Lieutenant General
1795	12/26	Major General Carl Friedrich von Klinckowström, later Lieutenant General, Count since 1798
1799	11/5	Colonel Johann von Malschitzky, later Major General

Regiments No. 28 and 29, organized in Wesel in 1723, were the first to be composed not of companies and battalions but of men transferred from all the regiments. In the regiment formed on January 3 for Major General Konrad Wilhelm von der Mosel there were "smaller and middle-sized people", because it came into being as a fusilier regiment, differing only in having fusilier caps. These regiments were intended for use in second lines and supplied with shorter guns. I.R. 15 was transferred to the Prignitz. Wesel was Prussia's western fortress in the Duchy of Cleves, where the Lippe flows into the Rhine. On April 1, 1727 it received men from I.R. 3, but in 1729 it gave men to I.R. 31. In 1739 it consisted of 1533 men and 58 supernumeraries. According to the regulation of February 28, 1714, the regiment was only present in full from April to June, being incomplete the rest of the year because of extensive furloughs. The law calling for compulsory service, replacing the previous volunteer system, came into being on May 1, 1733, though somewhat limited by exceptions. But since the King was seen as a soldier and the nobility was not excepted, the soldier's esteem and military pride increased when based on equal obligations. Service brought many a rise in social standing. When in 1740 the Bishop of Liege occupied the Barony of Herstal, which was offered to him for sale, the King, in Wesel, made a threatening demand on the subject of homage and within ten days sent a detachment of Westphalian and Wesel regiments under General von Borck, who collected contributions. On October 20 the Bishop paid 200,000 Taler for withdrawal. It was recognized in Europe that one did not take the new king lightly.

The regiment stayed in the west until 1741. On December 28 the King had Major General Count zu Dohna, its chief since 1733, transferred to Regiment 44, and the regiment, under Colonel de Hautcharmoy, made into a musketeer regiment and transferred to Brieg, the eastern fortress north of the Oder. Riedesel's I.R. 41 took its place in Wesel. The recruiting area for I.R. 28 was now the districts of Brieg, Ohlau and Kreuzburg with their cities. This move into the not yet conquered province on the edge of Upper Silesia, whose Catholics were very intolerant, was not good for the regiment. In February of 1742 it joined the Old Dessauer's Observation Corps in Upper Silesia. The infantry instructions of February 1, 1743 and the new regulation of June 1 called for "ceaseless progress" and said "that nothing matters but forcing the enemy to withdraw from the place where he is standing". In 1744 I.R. 28 was with the Second Corps under General von der Marwitz at Troppau and Jägerndorf, until Prince Dietrich led it via Cosel, Brieg and Grottkau to Neisse. In 1745 it again cleaned out Upper Silesia under Margrave Carl. The King knew: "Of all the means at my disposal, war is the only one that suits me". On June 4 I.R. 28 fought in the second encounter at Hohenfriedeberg. Under Lieutenant General Count von Waldburg it fought under heavy fire on the Saxon right wing in the Prince of Braunschweig's Brigade. At the end of June it mopped up Upper Silesia again, helped to take Cosel on September 5 and fought at Jägerndorf in mid-October.

In 1756 it belonged to the Silesian Corps under Field Marshal von Schwerin, holding enemy forces away from the County of Glatz from mid-September to the end of October so as to ease the Bohemian offensive. The four-column march over the mountains toward Jungbunzlau-Leitmeritz took place in mid-April. At Prague on May 6 it advanced with the left wing of the first line at Unterpoczernitz, coming into battle at the moment when the King turned the tide of battle with a new strong point. In Hautcharmoy's Division—he had been its chief for fifteen years—it led the breakthrough south of Kej against the unsupported Austrian main line and got into the bitter fighting in the Maleschitz Gorge. In addition to Schwerin and three other generals, Lieutenant General de Hautcharmoy fell here.

The regiment had heavy losses. Treskow took command of the division. The grenadiers shed blood on August 14 in the defeat at Landeshut; the regiment was captured by the Austrians on November 12 in the first siege of Schweidnitz. On April 16, 1758 the grenadiers, after a fourteen-week siege, stormed the city again and took the Gallows Fort, while the regiment could be reconstituted with more than 12,000 Prussians in the spring as a result of a prisoner exchange. The Silesian replacements had returned anyway after the King's intervention and the victory at Leuthen. At the beginning of May I.R. 28 marched with the King's Army to besiege Olmütz. There it was attacked at Domstadtl on June 30 while guarding the train of 4000 wagons from Troppau under Lieutenant Colonel von der Mosel, and decimated in a hard fight. When the siege was ended, Margrave Carl took command, and under him it marched to Spremberg in September and back to Silesia at the end of October. In 1759 it was camped at Schmottseiffen with the King, then it moved between Landeshut and Glatz under Fouqué, which the grenadiers were in charge of securing, and carried out the action at Konradswalde southwest of Landeshut. On June 23, 1760 the grenadiers fought bravely but were beaten at Landeshut. The regiment finally fought on the Warthe in Prince Heinrich's Corps, and between Glogau and Liegnitz from August 1 on. In 1761 it stayed in camp with the King at Bunzelwitz. The grenadiers went to Pomerania under Platen and came through the battles of Körlin, Spie and Kolberg with heavy losses. In the winter of 1761-1762 the regiment had to be reorganized, and in 1762 it saw no more action in Silesia, while the grenadiers, in Bevern's Corps, fought only at Reichenbach on August 16.

Despite its few successes, it was favored by the King after 1763. On the other hand, he grumbled on September 7, 1784: "The regiment is in such disorder that I shall have to send an officer from my regiment to it!" With the main army in 1806, it was disbanded in the Magdeburg capitulation, the Third Battalion by that of Brieg. Those on garrison duty joined the First Silesian Regiment, later I.R. 11.

Infanterie-Regiment Nr. 28

INFANTRY REGIMENT NO. 29
Regiment on Foot

Commanders of the Regiment

1723	8/14	Colonel Hans Christoph von Bardeleben, later Lieutenant General
1736	5/23	Colonel Georg Heinrich von Borcke, later Lieutenant General
1747	4/13	Major General Hans Caspar Ernst von Schultze, later Lieutenant General, died 12/2/1757 of wounds at Breslau
1758	1/5	Major General Carl Heinrich von Wedell, transferred to I.R. 26
1758	4/3	Major General Carl Gottfried von Knobloch, later Commandant of Schweidnitz
1764	5/22	Major General Johann Ferdinand von Stechow, later Lieutenant General
1778	3/23	Major General Heinrich Ludwig von Flemming, later Lieutenant General and Commandant of Breslau
1782	5/20	Major General Balthasar Ludwig Christian von Wendessen, later Lieutenant General and Governor of Warsaw
1798	3/1	Colonel Carl von Treunfels, later Major General

The sister regiment of I.R. 28 was organized in Wesel in 1723 of transfers from all the Markish and some East Prussian field regiments, under Colonel von Bardeleben, who took command on August 14. Again it was at first a fusilier regiment. As of 1716 the division among small towns had stopped; now a battalion or regiment was concentrated in one or a few locations for decades, complete with barracks, so that the regiments took on a regional character. On April 1, 1727 it, like I.R. 28, contributed a company with 220 supernumeraries to the formation of I.R. 31 in Wesel. The training time of almost twenty years was characterized by strict discipline, solidly grounded in the regimental commander. His decisions were examined by the Auditor General and reported to the King, who possessed a strong sense of justice. The punishments that seem incomprehensible today were then no different from those in civilian life. Through all the hard times and changes of commanders, locations, types of arms and service and troop spirit, the free will of the lower ranks was bound and punished. Instead, "desire and love for service" were to be striven for, and training was to proceed "without the slightest brutality"; at least that was the goal that the Old Dessauer and many others promoted. Until 1740 Wesel remained its garrison, and recruits came from the Duchy of Cleves.

With I.R. 28 it took part in the proceedings against Herstal from September 14 to October 20, 1740, after having been changed into a musketeer regiment during the summer—in other words, changing its hats. Until early April 1741 it remained in Wesel, then marched to join the Old Dessauer's Observation Corps in its camp at Göttin on the Plane, south of Brandenburg. Moved to Grüningen on September 12, it was relieved on September 12, and spent a year on drill and guard duty, wintering in Berlin. In 1742 I.R. 29 was to go to the Jägerndorf area at the end of April and advance each of the March. But the King turned toward Bohemia, where it came into battle at Chotusitz on May 17; I.R. 29 fought in the second encounter on the left flank just south of town, where it was hit by strong forces and heavy artillery fire. The bitter street fighting, in which the Austrian infantry held its ground stubbornly, made possible the final attack of the right wing with 21 fresh battalions. Colonel vonKreytzen and two grenadier captains received the Pour-le-merite. Peace was made barely four weeks later. The regiment stayed in Silesia and was garrisoned in Breslau. Recruits came from the districts of Liegnitz, Lüben and Trebnitz and their cities. In 1744 it was in Upper Silesia with von der Marwitz's Second Corps, in 1745 under Nassau and Margrave Carl. On May 22 it took part in the dangerous march back from Jägerndorf to Neustadt, during which the Württemberg Dragoon Regiment so distinguished itself at Bratsch in a counterattack on the Hungarian infantry. At Hohenfriedeberg on June 4 it helped to lead the first attack of the second line against the Saxon right wing, which threw them out of the order of battle. Major Adam Henning von Kamecke received the Pour-le-merite in June of 1747.

In 1756 it belonged to the Second Silesian Corps under Field Marshal von Schwerin and carried out limited attacks between Nachod and Aujezd (east of Olmtz), but at the end of October it withdrew. In April of 1757 I.R. 29 marched via Brandeis to join the King's Army, a remarkable achievement in leadership as a combined advance. After the failure of Winterfeldt's attack with Schwerin's and three Silesian regiments at Prague on May 6, it joined Hautcharmoy's Division to lead the breakthrough south of Kej. At the Rokenitz Brook the regiment suffered terrible losses until the King could redirect Schwerin's wing. At Kolin on June 18, five weeks later, it had the task of taking the decisive Krzeczhorz Heights on the left flank of the avant-garde under Major General von Hülsen, a move on which everything depended. Repeated attacks succeeded until 5:30 P.M., when the battle was almost won; then came counterattacks in an attempt to hold it, until all was lost. From now on luck deserted the regiment.

At the beginning of August the regiment, in Bevern's Corps, took a fortified position on the Lohe, which could be attacked only on November 22, after the fall of Schweidnitz and the victory at Rossbach, and then only after a four-hour artillery barrage. In the midst of this Lieutenant General von Schultze fell, a brave man, befriended by the King, and the chief of the regiment since 1747. Despite the King's advance, Breslau surrendered after five cannon shots. The occupation by the regiment reduced it to about one-tenth, which came to Berlin to be reconstituted. In 1758 I.R. 29 joined Prince Heinrich's Saxon Corps, which did a good job of protecting Saxony against superior enemy forces. Major General C. H. von Wedell was its chief for three months, until he was transferred to I.R. 26. In the spring of 1759 it took part in the first advance into Franconia, but then had to go on to Kunersdorf, where on August 21 it suffered heavy losses in the infantry battle on the Kuhgrund and the Deep Path. One battalion was taken prisoner at Maxen on November 21; thus the remeining battalion again occupied Breslau in 1760, while the grenadiers fought at Liegnitz and Torgau. In 1761 I.R. 29, at full strength again, joined the Goltz Corps and experienced the advance through Gostyn and Krölin and the breakthrough across the Spie Brook to the Württemberg Corps on October 2. When lines back to the Oder were opened, Major General von Knobloch, the regiment's chief since the spring of 1758, had to surrender the regiment in Treptow on October 25. In 1762 it got a rest at the Fortress of Breslau. Then the regiment and its grenadiers were called to take part in the siege of Schweidnitz from August 8 to October 10; possession of the city was decisive in controlling two-thirds of Silesia.

In Rüchel's Corps in 1806, it was disbanded in the capitulation of Pasewalk, the Third Battalion in that of Breslau. Those on garrison duty joined the Second West Prussian Regiment, later I.R. 7.

Infanterie-Regiment
Nr. 29

INFANTRY REGIMENT NO. 30
Regiment on Foot

Commanders of the Regiment

1728	2/18	Colonel Martin von Thile
1733	2/14	Colonel Joachim Friedrich von Jeetze, later Field Marshal
1752	9/23	Major General Christian Gottfried (von) Uchlander
1755	11/5	Major General Berend Sigismund von Blanckensee
1756	10/19	Major General Hans Samuel von Pritz
1757	1/21	Major General Ernst Ludwig von Kannacher
1759	1/1	Major General Joachim Friedrich von Stutterheim, later Lieutenant General
1768	11/23	Colonel Carl Franz, Baron von Sobeck, later Major General
1778	11/7	Colonel Philipp Wolfgang Teuffel von Birckensee
1782	5/20	Major General Georg August von Schönfeldt, later Lieutenant General
1792	11/7	Colonel Georg Friedrich von Wegnern, later Major General, died 9/29/1793 at Ensheim
1793	10/2	Major General Ernst Friedrich Wilhelm von Rüchel, transferred to the Guard Regiment in 1798
1798	1/13	Colonel Friedrich Adrian von Borcke, later Lieutenant General

As ordered on February 18, 1728, the Fusilier Regiment was founded in Anklam and Demmin for Colonel Martin von Thile. The qualified men of von Rinsch's Garrison Regiment, who came from Küstrin, Peitz and Driesen, formed the core of it, as well as units from the garrison companies of Spandau (since 1679) and Oderberg (since 1693) and supernumeraries of the field regiments, usually unranked because of "smallness". It became the best of the four fusilier regiments of King Friedrich Wilhelm I! Mobilized against Hannover in 1729, it went to the upper Rhine at the end of April 1734 in the War of the Polish Succession. Late in November of 1735 it returned to its garrisons. For nineteen years, beginning in 1733, its chief was Joachim Friedrich von Jeetze, whom the King valued highly and promoted to Field Marshal in 1747. Until 1740 positions as officers were also open to commoners. When they asked for titles of nobility, the King sometimes replied: "His sword ennobles him without this." Count Schmettau reports: "He was often pleased with the good progress of his officers, their diligence and the great sense of honor that inspired them. Then he asked him if he had ever seen in any other army a corps of officers with such a spirit." Many bourgeois officers earned a title of nobility. Only after 1763 was the precedence for the nobility made stricter. The garrisons were Anclam and, for one battalion, Demmin until 1796, then Stettin plus Kolberg for three companies. Recruits came from the districts of Anklam, Demmin, Treptow, Usedom and Wollin with their cities.

As a musketeer regiment it took part in the march to Silesiaon December 16, 1740, on the left flank of the First Corps. Its commander was Colonel Johann Georg von Lestwitz, who became chief of I.R. 31 in 1746. This flank encircled Glog auto meet the Second Corps and complete the circle. Then it went to Breslau, and via Ohlau to the Fortress of Neisse, which was invested from January 19 to 22, 1741. The Oder had to be kept free as a supply route. After the fall of Neisse on October 31, the regiment marched to northern Bohemia under Hereditary Prince Leopold. In February of 1742 the First Battalion was sent to northern Moravia, but came back to the Chrudim camp and joined the King early in May. At Chotusitz on May 17 it fought on the right wing, whose turning finally brought victory. The King was thrilled by the infantry: "Our infantrymen are nothing but Caesars!" In 1744 it was with the First Corps of the King, who led an unfortunate campaign. On February 14, 1745 the First Battalion stormed the Plomnitz Heights at Habelschwerdt. On June 4 at Hohenfriedeberg it took part in an attack on the Saxons in the second line, at Günthersdorf it fought against the Austrians. Lieutenant General von Lehwaldt led I.R. 30 with his corps to meet the Old Dessauer on December 13. At Kesselsdorf two days later it stormed the burning village from the northeast and captured 20 cannon, four mortars, a flag and numerous drums. Its commander, Lieutenant Colonel Ernst Ludwig von Kannacher, who became the regimental chief in 1757, and the battalion commanders von Kracht and von Horn received the Pour-le-merite.

In 1756 I.R. 30 was with the King's Army. While the grenadiers surrounded the Saxons at Pirna, the regiment fought at Lobositz on October 1, between the Lobosch and Homolka Mountains. When the left flank stormed down from Lobosch Mountain, it closed ranks along the road to Lobositz at Keith's command. It lost 275 men. At Prague on May 6, 1757 it served under Duke Ferdinand of Braunschweig as second reinforcement for the Hautcharmoy Division that broke through south of Kej. Three days later, the grenadiers and two companies took the Ziska-Berg northeast of Prague. The regiment was spared Kolin. In mid-August the King took the grenadiers along to Rossbach, while the regiment went to Silesia with the Duke of Bevern. Only the keen intellect of the King could master a defense at such distant points. After the opponents had faced each other across the Lohe for almost two months, defeat came on November 22. The regiment offered brave opposition. On December 2 it moved along to Leuthen; in the first line of the right wing it advanced over the Kaulbusch to the heights northeast of Sagschütz and the village of Leuthen. Then it helped to enclose Breslau, which fell on December 19. After the unsuccessful 1758 campaign in Moravia, I.R. 30 stayed in Silesia under Margrave Carl. At Hochkirch on October 14 Field Marshal Keith led the regiment along the eastern edge of the town against the lost batteries, and was knocked from his horse by a bullet. Only the specific command to leave immovable guns alone and to withdraw caused the regiment to give in. It had lost half its soldiers. The bitter "end of a wounded lion" prevented a total catastrophe. In 1759 the regiment fought at Torgau, Meissen and Böhmisch-Friedland with success. Under Prince Heinrich in 1760 it fought at Strehla on August 20 and Torgau on November 3, forming the avant-garde at the latter. From 1759 on it stayed at Alt-Stutterheim, while the grenadiers were at Lubath as a battalion. Together they led the second attack against the Süptitz Heights; after the breakthrough it was pressed and attacked on two sides by four regiments of cuirassiers. There were frightful losses; almost all the officers were wounded. In 1761 I.R. 30 helped to secure eastern Saxony, as also in 1762 until October 29, when it went into battle at Freiberg, where it advanced toward the northern part of the Spittel Forest in the Jung-Stutterheim Brigade and took several battlements, while its commander led the third attack column. Two majors and a lieutenant received the Pour-le-merite.

In 1806 it was first in Weimar's, then in Blücher's Corps, and was disbanded in the capitulation of Ratekau on November 7. The Third Battalion became the Second Kolberg Infantry Regiment when it occupied Kolberg in 1808; in 1813 it was divided into first and second units; in 1914 it became the Grenadier Regiment (Second Pomeranian) No. 9 Kolberg. The men on garrison duty stayed with it.

Infanterie-Regiment
Nr. 30

INFANTRY REGIMENT NO. 31
Regiment on Foot

Commanders of the Regiment

1729	2/1	Colonel Friedrich Wilhelm von Dossow, later Field Marshal
1743	8/1	Colonel Friedrich Wilhelm, Marquis de Varenne
1744	12/9	Colonel Philipp Bogislav von Schwerin, later Major General, transferred to I.R. 13
1746	2/12	Major General Johann Georg von Lestwitz, later Lieutenant General and Commander of I.R. 32
1763	6/25	Lieutenant General Bogislav Friedrich von Tauentzien, later General of the Infantry, Governor in Breslau
1791	5/20	Major General Philipp Ernst von Borck
1792	5/20	Colonel Philipp Friedrich Lebrecht von Lattorff, later Lieutenant General, Governor of Küstrin
1800	5/20	Colonel Bernhard Vollrad von Oldenburg, later Major General
1805	7/16	Major General Heinrich von Kropff

With the capitulation of February 1, 1729 the regiment was formed in Wesel for Colonel Friedrich Wilhelm von Dossow out of three "little battalions", each of 440 men, with "supernumerary additions" from the garrisons to provide the reserve manpower. The older Regiments 9 and 10 each had such a battalion, the newer I.R. 28 and 29 together provided one. Some of the men had already been trained in artillery. Dossow, promoted to Field Marshal in 1745, was a talented soldier with training in military science at the Colberg Knights' Academy from 1680 to 1683. The new regiment was the fifth in Wesel and served to strengthen the forces in the west, where Friedrich Wilhelm I had turned all garrison units into field troops. All in all, the army was more than doubled between February 1713 and June 1739. The basis of the new structure was the service time of four years for junior officers. One was to be "a real soldier with ambition" and could, after twelve years' service "and with very great merits and an open mind", as well as a good exterior, be promoted to Second Lieutenant, thus superseding the rank of Ensign. Ordinary soldiers were called "du", non-commissioned officers "Er", sergeants the courtly "Sie". In principle, the non-commissioned officers served for fifteen years or until they were no longer able to serve in the field. Many "merchants' and mayors' sons" hurried into this career, often moving into the government later after proving their capability. The King said: "The non-commissioned officers are the best feature of the company, and on the basis of experience, the older the better". From 1729 to July of 1743 the regiment was stationed in Wesel; it had been a musketeer troop since January 26, 1743, and was sent to Breslau. When von Dossow became chief of Fusilier Regiment 45, he was given a cadre of one hundred men. His successor was Major General J. G. von Lestwitz, punished with imprisonment by the King for the surrender of Breslau on November 23, 1757, but without losing his regiment, and in fact being given I.R. 32 in 1763. He was followed by General Friedrich Bogislav von Tauentzien, previously Commander of the First Guard Battalion and defender of Breslau in 1760, since 1763 Inspector of the seventeen Silesian infantry regiments, the artillery and engineers as well as Governor in Breslau. He was esteemed by the King, but less so by the people. On January 9, 1796 the regiment went to Warsaw, with three companies going to Lencziz. In 1788 the Third Battalion was created in Warsaw. Recruits came from the districts of Breslau and Namslau with their cities; Breslau itself was free.

After taking part in the move against Herstal from September 14 to October 20, 1740, the regiment stayed at home in Wesel during the First Silesian War. In 1744 it moved from Breslau to the King's First Corps, occupied Prague from September 11 to 16, and after the misfortunes in Bohemia it endured the unplanned three-column retreat to the Prussian border until early December. "The fruits of the field were fully consumed", desertion and discouragement increased. In 1745 it was sent with the Silesian garrison regiments and eight field regiments to occupy the fortresses. At Hohenfriedeberg the grenadiers surrounded the Saxon left flank at Pilgramshain; at Soor they provided security west of Staudenz.

In 1756 the regiment belonged to the Second Corps of Field Marshal von Schwerin in Silesia, which was limited to relieving action. In 1757 he ordered I.R. 31 to advance to Bohemia and unite with the King northeast of Prague on May 6. The regiment went into its first battle, along with the grenadiers; under Winterfeldt it advanceded up the Homolle Heights on the left flank at Sterbohol without firing. "Some came within 200 paces of the enemy and lowered their guns, some of the enemy turned, but the enemy fire felled the attackers by the hundreds". The battle line broke and began to fall back, the enemy attacked, until the second encounter when Zieten moved in. The regiment lost one-third of its men, the grenadiers twice that. The Silesian replacements fought here against their countrymen. Finally the regiment besieged Prague. After Kolin it marched to Silesia in July under the Duke of Bevern. Holding a position on the Lohe was meant to safeguard Breslau, protect Schweidnitz, win time for the King to return and maintain the army, although 19,000 men were just 9000 meters away. The center was commanded by Lestwitz, who was supposed to control the panic in Breslau after the breakthrough. The "free departure" of the regiments after the capitulation of November 25 was devastating. In 1758 it belonged to the Saxon Corps that advanced into Franconia. Hülsen led the regiment and grenadiers to join the Pomeranian Corps on the Warthe in mid-June. On July 23 it unsuccessfully stormed the Russian position at Kay. The King took it along to Kunersdorf on August 12. After the infantry battle in the Kuhgrund and the flight of the cavalry, he tried to use the regiment to build a line to protect the retreat. It lost 431 soldiers. The King "had every private paid eight Groschen because of his courage". In 1760 he took it again to the unsuccessful siege of Dresden, then to Ligenitz on August 15, where only the grenadiers came into battle and two commanders received the Pour-le-merite. On September 17 it was on hand for the storming of the heights at Hoch-Giersdorf, and on November 3 it was at Torgau, serving in the Tettenborn Brigade of Zieten's Corps to cover the the village of Süptitz in a flank march, losing 200 men but opening a part to victory. With the King's Army again in 1761, the grenadiers climbed the Adelsbach Heights on July 4, lost 300 men, and successfully besieged Schweidnitz from August 8 to October 10. The regiment and its grenadiers were almost always together, and in the Seven Years' War they lost neither a flag nor a cannon. After 1757 it had regained the King's esteem and was favored after 1763.

In 1806 it was disbanded in the capitulation of Schweidnitz; the grenadiers still fought in East Prussia and joined I.R. 2 in 1808, later becoming the First Kaiser Franz Guard-Grenadier Regiment No. 2. The Third Battalion formed the First Silesian Regiment, in 1808 the First I.R. 11, in 1914 the Grenadier Regiment King Friedrich Wilhelm II (First Silesian) No. 10 Schweidnitz.

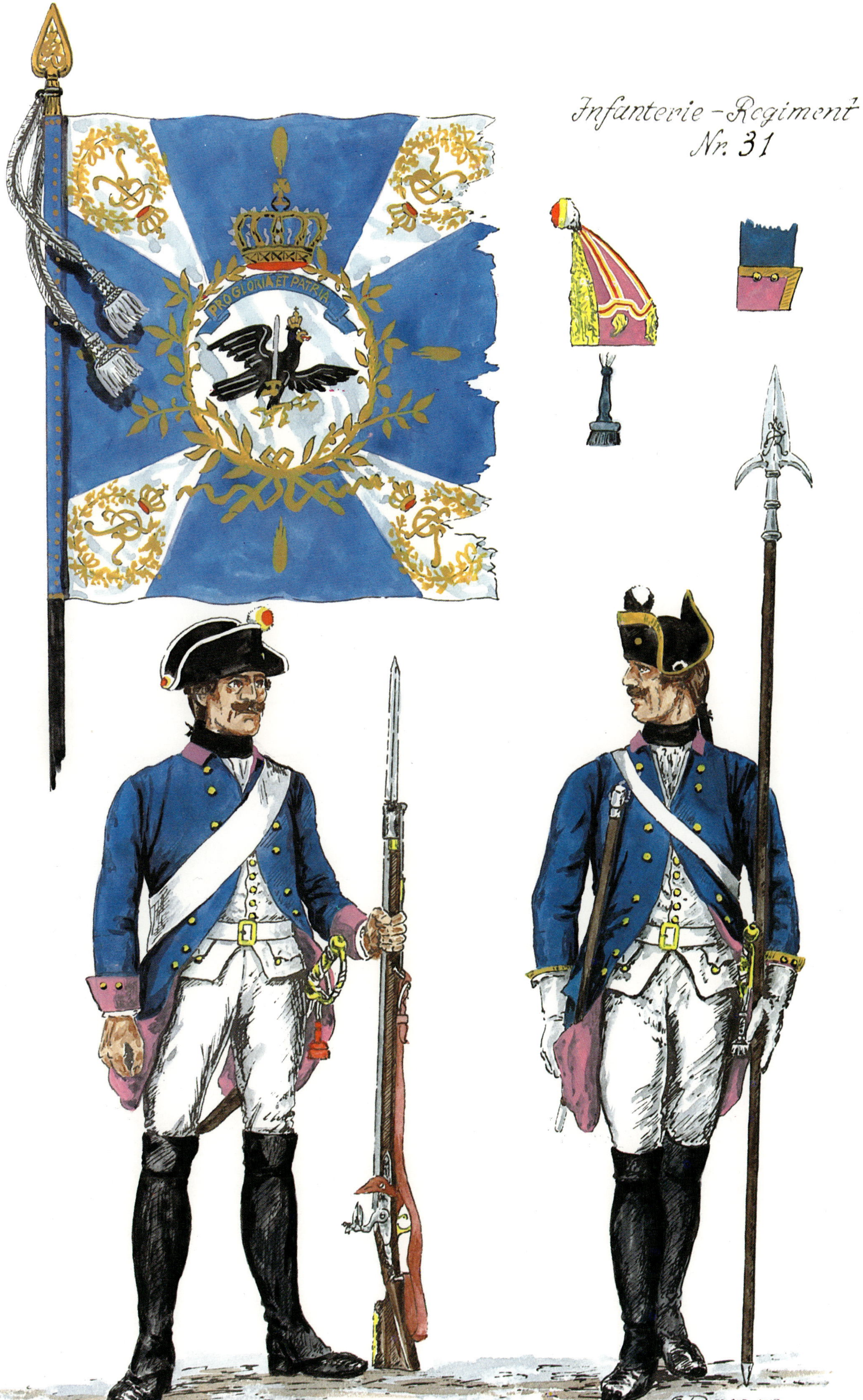
Infanterie-Regiment
Nr. 31
PRO GLORIA ET PATRIA

INFANTRY REGIMENT NO. 32
Regiment on Foot

Commanders of the Regiment

1743	8/2	Colonel Friedrich Julius von Schwerin, later Major General
1747	5/17	Major General Joachim Christian von Tresckow, later Lieutenant General
1763	6/25	Lieutenant General Johann Georg von Lestwitz
1767-	1770	vacant
1770	9/13	Colonel Hans Christoph von Rothkirch, later Lieutenant General and Commandant in Neisse
1786	3/1	Major General Friedrich Ludwig, Hereditary Prince of Hohenlohe-Ingelfingen, later General of the Infantry

On August 2, 1743 Colonel von Schwerin took command of the newly established regiment that had been transferred from Minden to Silesia that July, and that was called "Jung-Schwerin" to differentiate it from I.R. 24 and 31. Fusiliers at first, they were changed to musketeers in 1745 because more recruits were thus available to them. Since October 23, 1718 the First Battalion had been the garrison battalion in Geldern under Major General von Lilien, as of 1725 von Röseler, as of 1738 von Kröcker, and as ofNovember 1, 1730 a fusilier battalion. The Second Battalion consisted since 1720 of the two garrison companies in Lippstadt, which were joined in 1725 by two marine companies from Emden, established in 1724, and the Tecklenburg Invalid Company, in existence since 1723. As of June of that year its commander was Major General von Raders, followed in 1731 by Lieutenant Colonel von Beaufort. On October 1, 1734 it became a fusilier battalion, and the marine companies were reassigned. Its garrison was Neisse, which was captured only on October 31, 1741 and then built into a fortress. As of May 30, 1743 "Fort Prussia" was built on the left bank of the Neisse, likewise Glatz, Cosel and Peitz. In November of 1741 the Engineer Regiment was established there, afterward the Garrison Regiment 8 from Minden, with one battalion coming in the summer of 1742, the other in 1743. Recruits came from the districts of Neisse and Münsterberg and their cities. The Catholic population of Upper Silesia was furiously opposed to Prussia. For that reason the reliability of the regiment was open to suspicion.

In 1744 it belonged, as did many Silesian regiments, to the Second Corps of General of the Infantry von der Marwitz, which secured Upper Silesia and kept the entrances to Moravia open. The planned advance on Olmütz, though, was not carried out. The corps was at Jägerndorf-Troppau by December and also secured the County of Glatz. As a result of the King's withdrawal from Bohemia, it had to march back to Neisse via Cosel, Brieg and Grottkau, losing about 140 deserters, one-tenth of its strength, in the process. There was always friction between troops and civilians, but the King intervened with strict orders. After the enrollment regulations of 1743 it was able to fill its vacancies. On January 9, 1745 it moved to Neustadt O/S under the Old Dessauer. Radun and Ratibor were taken early in February, and from March on northern Upper Silesia could be pacified. After again vacating Ratibor and Troppau, it took part on May 22 in Margrave Carl's breakthrough from Jägerndorf to Neustadt. After that it went to the Neisse Fortress, while the grenadiers, at Hohenfriedeberg on June 4, stormed Thomaswaldau with the Grenadier Brigade on the left wing. From the end of June on, Upper Silesia had to be cleaned up again under Lieutenant General von Nassau, and on September 5 Cosel had to be retaken. In mid-October I.R. 32 was between Jägerndorf and Troppau. Field Marshal von Buddenbrock was the commanding general in Silesia.

From 1747 to 1763 the regiment's chief was Joachim Christian von Tresckow, called "my good Tresckow" by the King, Division Commander in Prague in 1757, successful besieger of Schweidnitz and defender of Neisse in 1758. When pressure was put on his wife, who lived outside Neisse, to make Tresckow surrender, she abandoned her property and fled to the fortress.

In 1756 the regiment, in the Second Corps of Field Marshal von Schwerin, took part in the limited fortress attacks along the southern border of Silesia between mid-September and the end of October. In 1757 it crossed the Sudeten Mountains under Schwerin, crossed the Elbe and, on May 6, met the King's Army in the highlands of Prosek. After the failure of the left flank's first attack at Prague, the King made use of Hautcharmoy's promising attack in the first line on the same wing, including the regiment, to break through into the enemy center. When Hautcharmoy fell, Tresckow took command. In the Rokenitz Gorge it got into murderous fighting. Then it invested Prague (until June 19), thus missing Kolin. In the Duke of Bevern's Corps, it suffered losses at Moys on September 7 and marched to Breslau, which it evacuated on November 25 with just one battalion. It was said: "A lot of Upper Silesians are better surrounded in the fortress than taken on long marches", typical of a shortage of experience in their proper treatment! In the King's Army in 1758 it experienced the attack at Domstadtl on June 30, meant to stop the transportation of supplies from Troppau to Olmütz, after the grenadiers under Tresckow had captured the strong Fortress of Schweidnitz between the end of March and April 16. "In the course of seven highly changeable campaigns it never lacked pay, bread or fodder, very rarely vegetables and even more rarely meat." The Prussian portions of meat lured many turncoats. The King knew "that if you want to build an army, you have to start with its stomach, for that is the foundation of it". With the Pomeranian Corps in 1759 it suffered the bloody defeat of Kay, in which its commander, Colonel Burkhard Hartwig von Jagow, fell. Since it had only one battalion left, it secured the Oder crossings to Kunersdorf at Göritzfrom August 11 on. In mid-September the battalion, under Finck, reached Wunsch's Corps at Torgau. On Semtember 21 the enemy was beaten at Korbitz. The King remarked: "To what huge mistakes do we owe our salvation!" In 1760 and 1761—again with two battalions—it was stationed at the Fortress of Schweidnitz. Garrison regiments had been called to the field to replace it. On October 1, 1761 it was taken prisoner there by Laudon's trickery, while commanded by von Zastrow, whom the King had left there on his own with only a quarter of the needed occupying troops. In 1762 it was not reorganized, but when the prisoners returned in February of 1763 it was reformed and filled with von Horn's Regiment, formerly in Saxony under Prince Friedrich August. In 1784 the King grumbled: "But it doesn't amount to much". Still in all, it fought very well from 1792 to 1794.

In Hohenlohe's Corps in 1806, it was disbanded by the capitulations of Erfurt and Prenzlau, the Third Battalion by that of Breslau. Men on garrison duty went to the First Silesian Regiment, Grenadier Regiment 10 in 1914.

Infanterie-Regiment
Nr. 32
PRO GLORIA ET PATRIA
G. Dorn

INFANTRY REGIMENT NO. 33
Fusilier Regiment

Commanders of the Regiment

1740	6/26	Colonel Andreas Johann de Persod
1743	8/17	Major General Samuel von Schlichting
1743	10/30	Colonel Asmus Ehrenreich von Bredow
1744	12/31	Colonel Heinrich August, Baron de la Motte-Fouqué, later General of the Infantry
1774	6/13	Major General Georg Reinhold von Thadden, later Lieutenant General
1784	9/24	Major General Friedrich Wilhelm von Goetzen, later Lieutenant General and Governor of Glatz
1794	3/18	Major General Theodor Philipp von Pfau, died 7/13/1794 at the Schnzel
1804	9/24	Major General August Ludwig von Schierstedt
1804	12/31	Colonel Ludolf August Friedrich von Alvensleben, later Major General

On June 26, 1740 Colonel de Persod took command of the regiment that had just been formed of the garrison battalion that had been in Magdeburg since 1737 as the Second Battalion, and of five companies of the "New Garrison" of Magdeburg, transfers from the Anhalt Regiment, and a grenadier company as First Battalion. The garrison battalion had been founded by the four princes of Anhalt in 1734 for Imperial service on the upper Rhine and turned over to Prussia on September 1, 1736, whereupon Colonel von Wachholtz took command of it in Aken. The officers came from the entire army. I.R. 33 remained a fusilier regiment. Its recruiting area was composed of small cities around Magdeburg, such as Seehausen, Alsleben and others. From 1744 to 1774 the famous Fouqué (1698-1774) was its chief; he had grown up as a page at the court of the Old Dessauer, who admired him just as did the Soldier King, who called him "an honor-loving man and officer". Of consciously controlled manliness, without fear or reproach, an absolute man of honor and frankly pious, he joines the Rheinsberg circle of friends of Frederick the Great, who kept in touch with him through an almost daily exchange of letters and thoughts. The King called him "one of the bravest and smartest officers", a combination not often found.

Early in April of 1741 the regiment was already moving to the Göttin camp south of Brandenburg to join Prince Leopold of Anhalt-Dessau's Observation Corps, while the King was at the Battle of Mollwitz. Troops also remained behind for protection in Magdeburg, Minden and Wesel. The corps, moved to Grüningen on September 12, was broken up at the end of October. The regiments were supposed to fill their ranks with volunteers in Silesia to spare their own recruiting areas, but to remember "that they live in my land, not the enemy's". In the winter it was part of the occupying forces in Berlin. After the main army left for Bohemia, the King took a new corps, composed mainly of regiments from the Observation Corps and commanded by the Old Dessauer, to Jägerndorf, starting on March 13, 1742 and bound for Olmütz. Early in May the regiment went on to strengthen Oppeln, where the grenadiers arrived on June 3 to secure the city and guard shipping on the Oder. The Peace of Breslau on June 11 brought Silesia and the county of Glatz to Prussia, and twelve infantry regiments, eleven of cavalry and several garrison regiments were stationed there. I.R. 33 went to Glatz, staying until 1806, with three companies in Silberberg on the southern edge of the Eulengebirge, which the King built up massively into a fortress starting in 1763, as had been done from 1744 on in Schferberg and Puhu, northeast of Glatz and above the Neisse Valley. The County now supplied the regiment's recruits. In 1744 it joined vonder Marwitz's Second Corps, which was in Upper Silesia securing the Glatz-Jägerndorf-Troppau line until year's end. On its return via Brieg and Neisse, beginning on December 21, about 120 soldiers deserted, which was not at all common in the army. To be sure, Silesia became Prussian overnight, but its possession was still in dispute, loyalty to Austria had not yet died out in the Catholic area, and the use of Silesian soldiers against homeland and emperor was simply premature. On February 9, 1745 the corps, under Nassau, captured Ratibor and reported: "The infantry is of such good will they they forget all fatigue and show the greatest desire to fight". On May 22 I.R. 33, under Margrave Carl, broke through to Neustadt with the Jägerndorf magazine. At Hohenfriedeberg the grenadiers covered the left flank of the first line. At the end of October the King considered the campaign finished, and from that Christmas on there was a decade of peace. The King then wrote: "All young officers should should be held precisely to service and subordination, so that the kind of people our captains and staff officers are can be called on again".

In 1756 the regiment was with Schwerin's Silesian Corps, which carried out restraining actions between Nachod and Aujezd from mid-September to the end of October. In the entire Seven Years' War it experienced only two big battles: on the morning of May 6, 1757 Schwerin joined the King at Prague. In the first avant-garde attack, the regiment stormed the Homole Mountains on the left wing under Winterfeldt and was thrown back bloodily, but fought "uncommonly bravely" in the infantry battle, under Fouqué, losing its commander, Colonel Balthasar Friedrich, Baron von der Goltz, as well as a flag during its withdrawal. The report that units of the regiment went over to the enemy during the battle only appears once and is not to be valued too highly. Next it helped to besiege Prague. In 1758 it tried in vain from early May to July 1 to encircle Olmütz under the King, then from the beginning of August on it guarded Silesia under Margrave Carl, then under Fouqué. With the King again in 1759, it was at Neisse in April and at Landeshut from summer on, where on June 23, 1760 it fought bitterly for more than five hours to defend the fortified Bober Heights against two and a half times its strength, five corps under Laudon. Landeshut dominates the pass to Trautenau. The main attack on a rainy night was directed against the mountains east of town, where Colonel Baron von Rosen, chief of the regiment, was in command. A secondary attack was directed against the positions south of town, cutting the route back. Hill after hill had to be torn away from the Prussians despite repeated counterattacks. Whoever was not killed was captured. The corps lost 1927 dead and over 8000 wounded and captured, including a blood-covered Fouqué. Zieten led the shattered remnants to Breslau. Landeshut was plundered bestially. The King, deeply shocked, acknowledged: "He acted like a Roman!" An exchange of prisoners had to wait because of the few Catholics. Refilled in 1763 from the former Saxon regiment of Prince Maximilian, later von Plotho, it was criticized by the King in 1784: "Thadden does not represent the insignificant country battalion!" He was always well inclined to the grenadiers.

With the main army in 1806, it was disbanded in the capitulation of Magdeburg; the Third Battalion, organized in Silverberg in 1788, joined the Fourth and Fifth in 1808 to form the Second Silesian Regiment No. 12, in 1914 the Grenadier Regiment King Friedrich Wilhelm III (Second Silesian) No. 11 of Breslau.

Infanterie-Regiment Nr. 33

INFANTRY REGIMENT NO. 34
Regiment on Foot

Commander of the Regiment

1740 6/28 Colonel August Ferdinand, Prince of Prussia, later General of the Infantry

Immediately after he came to the throne, Friedrich II, in the course of reorganizing the army, gave his youngest brother the Second Battalion of his "Crown Prince" Regiment (I.R. 15) plus half of the I.R. 6 Bodyguard Company, to reorganize and command as of June 28, 1740. It was filled out with men from the garrison battalions No. 1 of Memel Hospital and No. 2 of Natalis Pillau, and from grenadier companies as well. The officers came from the Guards. After I.R. 15 was transferred to Potsdam, it took over the garrison at Ruppin and the recruiting area comprising the district of Ruppin and a part of the Priegnitz, with the cities of Ruppin, Nauen, Lindow and Rheinsberg. Later the grenadiers were located in Templin and three companies in Nauen. Its commander was Prince August Ferdinand of Prussia (1730-1813), a colonel at age ten, taught by the multitalented Christoph Ludwig von Stille, who proved himself as a cavalry commander at Hohenfriedeberg. A major general in 1756, personally brave but no troop leader, treated with respect but benign condescension, he was mustered out as unable to serve in 1758, as a Lieutenant General. He was more a friend of literature and music. Succeeding Margrave Carl, he was from 1762 to 1811 the Lord Master of the Order of St. John, which he tried his best to renew. From 1767 to 1782 he was occupied with colonizing the Warthe Breach, and in 1811, in Prussia's time of need, he turned all the Order's possessions over to the state. From then on the flags bore the white Cross of St. John. After Finck's surrender of Maxen in 1759 he repeatedly criticized the King. Prince Louis Ferdinand was his youngest son.

At the beginning of April 1741 I.R. 34 moved to the Gttin camp to join the Old Dessauer's Observation Corps, then to Grningen, until it disbanded at the end of October and was quartered in Berlin. From March 13, 1742 on it again went with Prince Leopold of Anhalt-Dessau to the Jägerndorf area to advance toward Moravia east of the March. But when the King left Moravia and marched to northern Bohemia to protect Upper Silesia, it went along. On May 13 the King's Army was camped at Chrudim. On May 17 the regiment took a position between Chotusitz and the Cirkwitz Pond when the Austrians attacked at 7:30. Amid the chaos of bitter fighting around Chotusitz, General von Wedell led the regiment out of the second encounter to support the struggling regiments until the twenty-one battalions on the right flank attacked from the southeast and decided the battle. His commander, Colonel Johann Jakob von Pritz, was killed. In 1744 the regiment remained in the Anhalt-Dessau Corps, which protected Berlin, later serving against Hannover and Saxony. The grenadiers were at Magdeburg, where they guarded the transport of heavy artillery across the Elbe to Leitmeritz on ships, then took part in the unfortunate occupation of Prague under von Einsiedel, but returned unscathed after having to leave their line of march. After the victories of Hohenfriedeberg, Soor and Kath. Hennersdorf the King pushed the Dessauer into a battle against the Saxons to attain peace. Until December 12 he felt out the enemy, to the displeasure of the impatient King, then at Kesselsdorf on December 15 he attacked the enemy's key position. Here on the left flank the regiment took part in the attack across the Bachgrund and the storming of the heights on the far side of it, which helped decide the victory.

Used sparingly in the Seven Years' War, it stood out when it fought. From 1756 on it belonged to the King's Army and helped to surround the Saxons at Pirna, beginning on September 10. Early in January 1757 the King had the companies of 38 regiments strengthened by thirty men apiece. From May 2 on, the regiment and the grenadiers helped to enclose the west side of Prague, avoiding the battles of Prague and Kolin. Prince Ferdinand was wounded. After the withdrawal from Bohemia it passed by Moys, where the grenadiers had suffered high losses on September 7, and marched to Glogau to take a position on the Lohe before Breslau. When the enemy took Schmiedefeld, Mochbern and Hfchen in the middle on November 22 and General von Schultze fell, Prince Ferdinand grabbed a flag and led the regiment forward into a bitter counterattack. Under Zieten, it joined the King, taking on new courage, "a certain feeling of solidity and confidence". I.R. 34 fought at Leuthen with the remaining battalion on the left wing of the second line, while the grenadiers protected the left flank in both encounters. In 1758 the regiment besieged Olmtz in vain. On June 30 von Zastrow's battalion defended the giant Wagon Castle near Domstadtl, "determined to fight to the last breath", so that some of the wagons were saved. In Silesia under Margrave Carl from August on, it went to join the King in Dresden on September 11 and was with Retzow's Corps as Hochkirch, keeping the way back open. In 1759 it was encamped at Schmottseiffen with the King, and was spared Kunersdorf. In 1760 the grenadiers fought to the end under Fouqu at Landeshut on June 23; the survivors were taken prisoner. But the regiment fought on the left wing at Liegnitz on August 15, joining Anhalt-Bernburg in a counter-attack, overrunning the enemy forces and driving them back to Bienowitz with astoundingly heavy losses. What remained of it then went to Breslau. The King praised its bravery by naming it in a proclamation, awarding five Pour-le-merite, plus a hundred gold pieces to every captain. In late October of 1761 Schenkendorff led it to add strength in Pomerania, where it took part in the luckless campaign until year's end. In 1762 it stood out in the siege and capture of Schweidnitz from August 8 to October 8: a captain and two lieutenants received the Pour-le-merite. In 1763 it consisted of 1230 natives, 127 Saxons and 566 foreigners.

From 1792 to August 8, 1803 Carl von Clausewitz served with I.R. 34, until he became adjutant to Prince August of Prussia, son of the regimental chief, in whose home he also met his future wife, Countess Marie von Brühl. In 1806 the regiment was dissolved in the capitulation of Pasewalk on October 29.

Infanterie-Regiment
Nr. 34

INFANTRY REGIMENT NO. 35
Fusilier Regiment

Commander of the Regiment

1740 6/27 Colonel Friedrich Heinrich Ludwig, Prince of Prussia. later General of the Infantry, died 8/3/1802; vacant thereafter

The sister regiment "of the House" to I.R. 34 was founded one day earlier, on June 27, 1740 as a fusilier regiment for the second brother of the new King Friedrich II, and made of helf the I.R. 6 Bodyguard Company. grenadiers from Garrison Battalions No. 1 of Memel and No. 2 of Pillau, and newly acquired volunteers from the Empire. No replacements were allowed to be taken from recruiting areas. The officers came from I.R. 6 and other field regiments. The soldiers were of smaller size and had shorter guns and fusilier caps. Their commander, Prince Heinrich of Prussia, born January 18, 1726—the twenty-fifth birthday of the kingdom—was a colonel at age 14. His father's favorite, the most talented brother of the King, who chose C. L. von Stille as his instructor, he had a difficult time as a handicapped rival in the King's shadow. He had honors heaped on him but was not allowed any influence. In 1744, for example, he was given Rheinsberg Castle. A field general in the classic style and the victor at Freiberg, he tendered his resignation in 1760 and 1762 because he regarded the King as a "gambler" out of opposition and ambition, and criticized him frankly. Until 1756 the regiment was stationed in Potsdam, from 1763 to 1795 in Spandau, with one battalion in Nauen, then in Königsberg/Neumark, the Second Battalion in Pyritz, the grenadiers in Soldin and three companies in Küstrin. It moved into the province at a slow pace. Like the Guards, it had no recruiting area, but was given thirty recruits a year from the Silesian "King's Canton", plus volunteer recruits. From 1795 on it drew recruits from the districts of Bomst, Meseritz, Deutsch-Krone, Posen and Hohensalza in West Prussia and Posen. It could not participate in the strengthening of the companies that began in 1755, and thus remained at the old head-count of 697 per battalion.

In the spring of 1741 the regiment moved into the Fortress of Magdeburg, where it stayed during the whole campaign. When the infantry received new instructions on February 1, 1743 as to "how drill shall be done in the future", the King had a different colonel drill the regiment: "I have found it appropriate, sir, to bring order into your regiment; it began to be lost". Other voices also mention certain disciplinary problems. Silesia had to contribute to the cost of the army, independently of the old provinces, and bear the cost of newly organized units itself from 1742 on. In the King's First Corps in 1744, it besieged Prague from September 11 to 16 and then was part of the occupying forces under Lieutenant General von Einsiedel. After the King's withdrawal, which began on November 22, the regiment made its way successfully through Leitmeritz, Böhm and Leipa, though it lost its artillery, followers and deserters. In 1745 it was stationed in Upper Silesia, then in a fortress, unused until the war was over. Prince Heinrich himself had done an outstanding job.

In 1756 I.R. 35 besieged the Saxons at Pirna from September 10 to October 16. The King had hoped to bring the Saxons over to his side in the war. On July 30, 1757 the regiment received reinforcements from the former Rochow Regiment while in camp at Bautzen on July 30, since on January 9 of that year it had not been eligible for the general increase in manpower. After the attack on the First Battalion at Hirschfeld on February 20, two captains received the Pour-le-merite. Early in April it joined Bevern's Corps at Reichenberg, where it was honored for the breakthrough with the Grenadiers' March. Decimated at Ostritz, the restored Second Battalion led the left attack under Bevern and got into the bloody battle by the Rokenitz Brook. The grenadiers were on the west side of Prague. Prince Heinrich gained the victory by his attack on the north wing. At Kolin under Treskow, it showed outstanding courage in the second line of the left wing, attacking the Kamajka Heights, led by the King with drawn sword. An hour and a half later its bayonet attack broke through the enemy battle line southwest of Krzeczhorz. But shortly before the victory, the Saxon cavalry attack fully scattered the First Battalion. Bevern led the survivors past Moys, where the hastily summoned grenadiers had arrived too late, to a position on the Lohe. On November 22 the Second Battalion, on the left wing under Zieten, twice regained the village of Kleinburg. At Leuthen on December 5 it again fought on the left wing of the second line, along with the grenadiers. Its achievements there founded the regiment's reputation. In 1758 it besieged Olmütz with the King from early May to July 1, without success. First Lieutenant von Lattorff won the Pour-le-merite here. In the attack on Domstadtl on June 30, the grenadiers safely convoyed one hundred pay and ammunition wagons through. It was too weakened for further action, as it had no recruiting area of its own and the Saxons did not fully provide replacements. In the summer of 1759 it was with the King again in camp at Schmottseiffen, restored to two battalions, and on August 12 it was at Kunersdorf with its grenadiers; repeated attacks on the batteries on the Great Spitzberg came to a stop at the Deep Path on account of exhaustion, heat and lack of ammunition. In the crisis the King—with two horses shot and bullet holes in his coat—took a flag and tried to stop the regiment from falling back: "I must do my duty here like any other!" Then the Austrian cavalry attack decided the outcome. The regiment was the last to leave the battlefield. At first in Prince Heinrich's Corps in 1760, it took the heights of Hoch-Giersdorf and seventeen cannon on September 17, led by Wied. At Torgau on November 3, it fought in the second encounter on the left flank with the Queiss Brigade, whose second attack on the Süptitz Heights came to a standstill. The cavalry attack from two sides caused it heavy losses. In 1761 it was camped at Bunzelwitz with the King. In 1762 it took part in the conquest of Schweidnitz from August 8 to October 10.

At the May review in 1764 the King praised his brother as "the only general who made no mistakes", although his brother stood before the regiment without a spontoon, because he regarded it as a demotion in rank. One year later Prince Heinrich bore his spontoon correctly in review, but then the King ordered him to give it up, sit on it and review the troops with him. In 1806 the regiment was disbanded in the surrender of Erfurt and Magdeburg on November 8.

Füsilier-Regiment
Nr. 35

INFANTRY REGIMENT NO. 36
Fusilier Regiment

Commanders of the Regiment

1740	6/24	Colonel Gustav Bogislav von Münchow, later Lieutenant General
1766	6/15	Colonel Henning Alexander von Kleist, later Lieutenant General and Governor in Spandau
1780	6/4	Major General Peter Christoph von Zitzewitz
1785	4/5	Colonel Wilhelm Magnus von Brünning (Brünneck), later Major General
1786	12/29	Major General Karl Albrecht Friedrich von Raumer, later Lieutenant General
1793	2/23	Oberst Georg Henning von Puttkamer, later Lieutenant General

This regiment was founded immediately after the accession of Friedrich II and taken command of on June 13, 1740 by Colonel von Münchow. It was composed of men capable of field service, taken from the garrison regiments of Küstrin, Spandau, Peitz and "Fort Prussia" near Stettin, plus grenadiers from the Hellermann Garrison Battalion No. 3, Kolberg, and volunteers recruited from the Empire. Most non-commissioned officers came from I.R. 6, the officers from many field regiments. The regiment was established in Potsdam but was overwhelmingly Markish. The cadres were formed by garrison units instead of field troops. Little is known of the man who commanded the regiment for 26 years. The establishment of new regiments offered officers better chances of promotion. For a time the King refrained from circumventing the strict principles of rank, without any disadvantages to the older regiments. Excessive harshness to the soldiers was decreased by the "regulation of how the commanders are to maintain the companies" of October 28, 1740« In all of this, the people are not to be treated cruelly or beaten, unless a fellow did not want to be reasoned with or persuaded with kindness, or the officer had ordered him several times and he had not obeyed". The garrison was in Brandenburg, with the grenadiers in Küslin. Recruits came from the districts of Osten and part of Greifenberg, Belgard, Borck and Polzin, with the cities of Treptow, Naugard, Massow, Wangerin, Wollin and Polzin in Pomerania. The city of Brandenburg was excluded.

In the spring of 1741 it went to Silesia with the Third Corps, intended for occupation duty but not yet for battle, with the First Battalion sent to Breslau, the Second for the Fortress of Glogau as of March 1, and the grenadiers fighting there and at Mollwitz in the center of the second encounter. In Glogau Colonel von Schwerin commanded I.R. 36; meanwhile the garrison regiment was organized. At the beginning of February 1742 it was brought together and transferred to Upper Silesia in the Prince of Anhalt-Dessau's Corps, which was to move east of the March from Jägerndorf. Southern Upper Silesia guaranteed access to the Moravian Gates, where the grenadiers were already stationed. In the King's First Corps in 1744, it marched in mid-August to join the siege of Prague, which fell on September 16. Before he returned, the King strengthened the city's occupying forces under von Einsiedel, including the Second Battalion, which then had to fight its way through with difficulty. The grenadiers fought on October 9 as the rear guard at Moldauthein. From mid-April of 1745 I.R. 36 secured forts in Silesia. After Hohenfriedeberg, where the grenadiers fought on the right wing, it was put under Lieutenant General von Nassau at the end of June to mop up in Silesia.

In 1756 it marched with the King's Army along the Elbe to Lobositz, where on October 1 the First Battalion, between the grenadiers and musketeers of I.R. 27 on the left wing, came into a firefight with the Croats, until at noon a spontaneous attack took the town within an hour. It had distinguished itself. In 1757 it went with Schwerin's Army from Silesia to Jungbunzlau but then served under Keith to enclose Prague from the west. At Kolin on June 18 it joined Major General von Hülsen's avant-garde to lead the first attack against the Krzeczhorz Heights, took town and battlements, drew heavy enemy forces to attack it, cleaned out the Eichenbusch, attacked six more times in a storm and held its position to the bitter end. It lost 23 officers and 908 men, having only between 569 and 720 men by evening. In mid-August it marched with the Duke of Bevern through Moys to the fortified position on the Lohe west of Breslau. When on November 22 the Austrians finally began their attack with superior forces, it fought under Zieten on the left wing and took part in the counterattacks on Kleinburg, though it had scarcely the strength of one battalion. On December 2 it united with the King's Army after its victory at Rossbach. This meeting raised spirits and inspired confidence. The grenadiers and the remaining battalion fought at Leuthen, again on the right wing of the second line, achieving "miracles", as the King said. The commander and both captains of the regiment in Grenadier Battalion 35/36 received the Pour-le-merite for their fighting and capture of four cannon on November 22 at Kleinburg, and of one battery at Leuthen, where the battalion lost 167 men. At last the remnant of the regiment was sent to join the siege of Breslau.

From the end of March to April 16, 1758 it was with the King's Army successfully besieging the Fortress of Schweidnitz, and advancing to Olmütz until June 1 to capture the city of Littau to the northwest of Olmütz and hold it against eight times its strength. Then it occupied Silesia under Margrave Carl. In 1759 it was at the Schmottseiffen camp, first under the King and then Prince Heinrich, while the grenadiers fought at Kunersdorf. After fighting at Hoyerswerda on September 25 and Pretzsch on October 29, now in Finck's Corps, its misfortunes and loss of power began when the corps was encircled and devastated at Maxen on November 20-21. For hours the battle raged in the forest, as Hülsen, detained by heavy snow, arrived too late. Finck surrendered with 13,000 men. Its reformation in 1760 was difficult because the old regiments took all the recruits regardless. The remaining battalion, relieved of field duty, went to Schweidnitz. There in 1761 it was taken prisoner along with three other battalions. It was not reformed in 1762, but only in 1763, when its ranks were filled from the former Saxon regiment of Prince Lubomirski, later of Röbel.

With the main army in 1806, it was disbanded on November 8 in the capitulation of Magdeburg, including the Third Battalion. The grenadiers joined Kolberg's Waldenfels Battalion, in 1914 the First Guard-Grenadier Regiment No. 1, while the men on garrison duty joined the First Pomeranian Regiment.

Infanterie-Regiment
Nr. 36

INFANTRY REGIMENT NO. 37
Fusilier Regiment

Commanders of the Regiment

1740	6/25	Colonel Paul Heinrich Tilio de Camas
1741	4/14	Colonel Peter Ludwig du Moulin, later Lieutenant General
1755	9/12	Major General Heinrich Adolf von Kurssell, died 9/26/1758 of wounds at Zorndorf
1758	12/8	Major General August Wilhelm von Braun, later Lieutenant General
1770	7/1	Colonel Johann Georg Wilhelm, Baron von Keller, later Lieutenant General, Commander of I.R. 8 and Governor in Stettin
1785	4/5	Colonel Adam Heinrich von Wolframsdorf, later Lieutenant General
1794	12/29	Major General Johann Rudolf Hiller von Gärtringen
1799	2/26	Colonel Johann Friedrich Gustav von Stockhausen, later Major General
1804	9/24	Colonel Carl von Tschepe, later Major General

Two days after the funeral of King Friedrich Wilhelm I, Colonel de Camas took command of the newly formed regiment, composed of capable soldiers from the Kolberg Garrison Regiment No. 3, under Hellermann in the fall of 1740, with non-commissioned officers from I.R. 6 and men recruited in the Empire, plus officers from the army and grenadiers from Pillau. It was established in Brandenburg/Havel, in an area which it had taken over from the "King's Regiment" I.R. 6. The recruitment of 'foreigners' had been made easier as of June 24, 1740—one day before its founding: "It no longer matters whether they are fleeing". It mattered only that the regiments be filled out with capable men. The expenses, which had gone up, were cut, but still amounted to between 30 and 200 Taler per man. Frenchmen were not accepted. As of August 8, 1740 monthly rosters were to be turned in which were divided between native sons and 'foreigners'. After the Peace of Breslau on July 11, 1742, I.R. 37 was sent to Glogau and given the districts of Glogau. Guhrau and Schwiebus, including their cities, as its recruitment area. In 1795 only the grenadiers remained in Glogau, as the regiment was transferred to Fraustadt und Lissa, with three companies going to Rawitsch. Now its recruits came from the districts of Schrimm, Schroda, Krotoschin, Posen and Kosten, including some of their cities.

Like I.R. 36, the regiment marched to Silesia at the end of the Third Corps in February of 1741 to do occupation duty, and occupied Breslau with the grenadiers. According to old rights, Breslau had garrison freedom, which the King recognized "only under the present conditions". The regiment was to prevent the Austrians from occupying the city. On August 10 the Breslau City Guard was absorbed to become the cadre of a new garrison regiment. In February of 1742 it went to Ratibor, while the grenadiers went to Moravia. Early in May Prince Leopold of Anhalt-Dessau took command in Upper Silesia. In 1744 I.R. 37 belonged the the King's First Corps and marched into Bohemia. From September 11 to 16 it helped to capture Prague. When Prague was cut off and the King began to think of retreating, the grenadiers fought in the rear guard at Moldauthein on October 9. The privations of the failed campaign depressed the army's mood; everyone was thankful when they reached the border early in December. In 1745, under Margrave Carl, it broke through from Jägerndorf to Neustadt. On June 4 it was in the center of the second line at Hohenfriedeberg, with the grenadiers on the right wing, which Lieutenant du Moulin, its chief, commanded and later turned toward the southwest. In the attack of eighteen battalions against thirty-three enemy battalions at Günthersdorf, the outcome was obvious in advance. Du Moulin, a proven corps leader, received the Order of the Black Eagle and pursued the enemy. At the end of June the regiment, under Nassau, went to Upper Silesia to mop up at last, and took Cosel on September 5. On September 30 the grenadiers conquered the dominating Graner Koppe at Soor in an attack of the second line.

The regiment belonged to Schwerin's Silesian Corps in 1756; after attacks that took it off its course, it reached Bohemia only in 1757 to meet the King northeast of Prague on May 6. On the left wing, it joined with the leading battalions in a vain attack on the Homole-Bergunder small-arms and artillery fire, suffering the heavy loss of thirteen officers and 661 men. Schwerin was killed and Winterfeldt wounded. The King intervened, leading the left wing to threaten the enemy's line of retreat. While Bohemia was being mopped up, I.R. 37 went with the Duke of Bevern to Silesia, where the Second Battalion went to strengthen Schweidnitz. On November 12 the Austrians took some outlying forts and shot breaches in the main fortifications. Then Major General von Sers surrendered. The grenadiers held Pilsnitz, before Breslau, on November 22, throwing back the enemy three times. At Leuthen on December 5 the First Battalion covered the left flank between the first and second lines; the grenadiers did the same on the right. The First Battalion captured three cannon. In winter the prisoners returned via Jägerndorf and Peterswald. With the King's Army in 1758, it besieged Schweidnitz until the storming of the Galgenberg on April 16, while the grenadiers were at Olmütz. When Friedrich left Bohemia early in August, it marched to the Oder to reinforce the Dohna Corps and fought at Zorndorf on August 25. Its commander, Major General von Kurssell, was mortally wounded there. After that it served under Prince Franz von Braunschweig at Beeskow, protecting Lusatia against the Austrians. The grenadiers remained in camp at Hochkirch in October 14. In Wedell's Corps, the regiment beat back the Swedes at Fehrbellin on September 22 and defeated the Imperial army on the Mulde at Torgau and Eilenburg on November 12. Getting recruits from Lower Silesia was no problem. In 1759 it fought under von Knobloch at Saalfeld on March 26, at Kronach early in May, and on July 30 it started toward Crossen with the King. On August 12 it was all but wiped out before the Great Spitzberg at Kunersdorf, losing sixteen officers and 992 men. On November 9 the grenadiers joined Finck's Corps and fell into captivity at Maxen. Their sense of honor and "spirit of goodness", amplified by the King, survived danger and suffering. On June 23, 1760 the First Battalion went down fighting under Fouqué at Landeshut; the Second was in Zieten's Corps at Freiburg. With Zieten at Wahlstatt and in camp at Bunzelwitz in August, it marched with Platen im September, advancing past Gostyn, Landsberg and Körlin to Kolberg, the siege of which was begun at Spie on October 2. On October 25 the Second Battalion was captured at Treptow, while the First endured the dreadful retreat from Kolberg to Stettin. In 1762 it fought in Silesia: at Adelsbach on July 6, at Friedland on July 8, and on July 21 it stormed the southern battlements of Leutmannsdorf up a coverless slope, and finally it besieged Schweidnitz for sixty-three days. In 1763 it consisted of 1485 Prussians, 27 Saxons and 342 'foreigners'! In 1784 the King grumbled about the regiment: "Keller resembles a heap of untrained farmers!"

In Rüchel's Corps in 1806, it was disbanded in the Magdeburg capitulation on November 8, the Third Battalion in that of Glogau.

Infanterie-Regiment
Nr. 37

INFANTRY REGIMENT NO. 38
Fusilier Regiment

Commanders of the Regiment

1740	6/23	Colonel Wilhelm Alexander, Count zu Dohna, later Lieutenant General
1749	8/24	Major General Johann Christoph von Brandes, later Lieutenant General
1758	12/20	Major General Carl Anton Leopold, Baron von Zastrow, later Lieutenant General
1766	9/5	Colonel Friedrich Gotthelf von Falkenhayn, later Lieutenant General and Governor in Schweidnitz
1781	7/4	Major General Heinrich Wilhelm von Anhalt, transferred to I.R. 2
1783	12/11	Major General Eberhard von Hager
1790	2/11	Major General August Wilhelm, Baron von Vittinghof, later Lieutenant General
1796	5/1	Colonel Christian August von der Marwitz, later Major General
1800	9/9	Colonel Carl von Pelchraim, later Major General

Colonel Wilhelm Alexander, Count zu Dohna, took command of the newly established regiment on June 23, 1740—it was the first day of Friedrich II's reign, one day after his father's burial. I.R. 38 was composed of field-capable soldiers of Garrison Battalion 1 of Memel, de l'Hospital, and 2 of Pillau, de Natalis, junior officers of I.R. 6, officers of other field regiments, and recruited 'foreigners' from the Empire. In February of 1741 it was ready and received its fusilier caps. In order to avoid confusion, it was called "Jung-Dohna"! When Austrian movements began to center around garrison-free Breslau, the King surprisingly ordered the inner city occupied on August 10 of that year, by a detachment under Field Marshal von Schwerin, including the regiment. Lieutenant General von der Marwitz was the governor. In 1742 I.R. 38 was sent to Neusalz, Neustädel near Beuthen/Oder and Glogau when the Peace of Breslau was signed on June 11 and the garrison regiment was in Breslau. In 1743 it was in Schweidnitz, from 1746 to 1756 in Liegnitz and Neumarkt. It was in Breslau again from 1763 to 1782, in Frankenstein from 1782 to 1796, and from then on in Neisse. Its location had shifted to the southeast. Its recruiting area was southern Upper Silesia with the districts of Leobschütz and Cosel, including their cities, a Catholic but very German-thinking area. The soldiers' training time lasted three years. The completion of the regiments was first rejected in Upper Silesia. For that reason General von der Marwitz agreed with the civil authorities on a yearly number of recruits. On August 16, 1743 the King issued the "Silesian Recruiting and Districting Regulations", according to which qualified men were to be examined in writing by a staff officer and captain of the regiment as well as deputies of the War and Domain Committees, though there were many exceptions. The lists were checked again by General von der Marwitz along with the committees. Until 1746 each regiment received up to eighty, afterward only up to forty Silesians annually.

In 1744 the regiment went with von der Marwitz's Corps to Upper Silesia, marching only to Jägerndorf and Troppau without going on to Olmütz. That was enough to protect the County of Glatz. From December 21 on it returned to Neisse via Cosel and Brieg, under Prince Dietrich. In January of 1745 it occupied southern Upper Silesia again, under the Old Dessauer, but left with Margrave Carl to break through from Jägerndorf to Neustadt on May 22 and march on to Hohenfriedeberg. There on June 4 it fought in the middle of the second line, which swung to the southwest to block the Saxon retreat line and attack the Austrians between Thomaswaldau and Günthersdorf. At the end of June it returned to Upper Silesia under Nassau, and retook Cosel on August 27 and Jägerndorf, Troppau and Ratibor by mid-October. In 1756 it was part of the King's Army that surrounded the Saxons at Pirna. The Saxons surrendered on October 16. On January 9, 1757 the companies were strengthened by thirty men each. At Landeshut the regiment was with Winterfeldt's Corps, which was supposed to support Schwerin's Corps. The destination was Jungbunzlau. On May 6 it joined the King northeast of Prague. Before that the grenadiers had taken the Elbe bridge at Altbunzlau and inflicted heavy losses on the enemy. In the Battle of Prague it was the third reinforcement in the right flank's second encounter; the King led it to join Lieutenant General von Hautcharmoy in the 28-battalion breakthrough south of Kej. Next it came into battle at the Rokenitz Brook. In the first attack of the left wing the grenadiers lost 278 men. At the end, the King led the left wing to victory on a wide front. The promising siege of Prague was wasted because of Kolin. After the failed Bohemian campaign it marched to Silesia with the Duke of Bevern. After Moys and the cannonade of Barschdorf, it was on the Lohe before Breslau from late September on; on November 22 it went into battle along with the grenadiers on the right wing under its chief, Lieutenant General von Brandes, "whose troops held their position most bravely until darkness". The regiment was "completely ruined", each battalion losing about 300 men, plus Colonel von Geiss. Before Bevern's night attack with the left wing took place, the regiment had left with the right wing for Breslau, which was evacuated on November 25 on terms of "free departure".In the process, nine-tenths of the soldiers ran away, believing Prussia was at an end. What remained of the regiment marched to Berlin to be reformed. The replacements it received in 1758 were mostly Markish, and some of the Upper Silesians found their way back after Leuthen. Now it belonged to the Saxon Corps, which advanced at Hof and Bamberg and then tried to secure Saxony at Zschopau, Pirna and Dresden. At the end of July 1759 the King took the corps to Bautzen; early in August the regiment followed him to Kunersdorf, where on August 12 it was in Finck's Corps on the right wing, opening the attack over the Trettin Heights and reaching the northwestern slopes of the range of hills. After the defeat its artillery held the pursuers off. Colonel von Steinwehr, the regimental commander, had fallen; losses had been high. After that one battalion fought in Wunsch's Corps in Saxony, at Torgau and Korbitz, and was captured at Maxen on November 21, under Finck. In 1760 its one battalion was in Schweidnitz, where on October 1 it was taken prisoner again. It was not reformed in 1762—prisoner exchange had stopped—but after the return of prisoners in the latter half of February 1763 it was reconstituted with the grenadier battalion of Rotkirch, formerly of Köller, which had been formed of Saxon grenadiers in 1757 and served in Pomerania since 1761.

In 1784 the King complained of the unfortunate regiment and its "wretched commander", meaning Major General von Hager. It served as a field regiment in 1807 and was disbanded in the surrender of Neisse. In 1808 the Third Battalion, after the successful defense of Cosel, joined the Second Silesian Infantry Regiment, in 1914 the Grenadier Regiment King Friedrich III (Second Silesian) No. 11.

Infanterie-Regiment
Nr. 38

INFANTRY REGIMENT NO. 39
Fusilier Regiment

Commanders of the Regiment

1740	6/29	Colonel Ferdinand, Prince of Braunschweig, later Field Marshal and Duke
1744	12/17	Prince Albrecht of Braunschweig, without rank because of his youth, died 9/30/1745 in storming the Graner Koppe at Soor
1745	12/26	Prince Friedrich Franz of Braunschweig, Colonelas of 3/1/1751, later Major General, died 10/14/1758 at Hochkirch
1758-	1763	vacant
1763	7/29	Colonel Wilhelm, Prince of Braunschweig, later Major General, died with the Russian Army in Bessarabia
1771	5/7	Major General Wichard Joachim Heinrich von Mollendorff, later Field Marshal
1783	12/11	Colonel Christian Ludwig von Knitz, later Lieutenant General
1793	3/23	Colonel Daniel Ludwig von Crousaz, later Major General

The first of the three following regiments from allied princely houses was supplied by the related House of Branuschweig for the King's brother-in-law and the reigning Duke's brother, Colonel Ferdinand, Prince of Braunschweig (1721-1792), who became its commander on June 29, 1740. The personnel came from the Sommerlatte Regiment, which had fought in the Emperor's service against the Turks in Hungary from 1737 to 1739. The non-commissioned officers were taken partly from foreign service, the officers from old field regiments. The grenadiers came from the garrison regiments of Memel and Pillau. Because of long negotiations, which even involved the Queen, the establishment at Templin took until July of 1741. At first the regiment was stationed in Stettin, then in Berlin as of August, and as of September 1741, when peace was made, in Magdeburg. For thirty years the regiment belonged to the House of Braunschweig, because Prince Ferdinand was succeeded by his brothers, of whom two died young at Soor and Hochkirch. Of Prince Friedrich Franz (1732-1758), in Prussian service only since 1744, the King said after his death: "With his talent he united determined bravery; this is hereditary in his family",high praise. Since 1755 it was called "Jung-Braunschweig" to distinguish it from I.R. 5. Until 1756 its garrisons were Königsberg/N, Soldin, Lippehne and Berlinchen, from 1763 to 1769 only Königsberg/N, from 1770 one battalion was in Soldin and the rest of the regiment in Königsberg except the grenadiers, who were in Pyritz. In 1795 it was transferred to Posen, with three companies and the grenadiers in Kalisch. Replacements came from free recruiting; it had no area, since the Dukes of Braunschweig were responsible for recruiting. In 1756 it had only 393 Prussians and 1261 'foreigners', mostly from Braunschweig. When their arrival decreased in the Seven Years' War, it repeatedly received limited recruits from the Silesian mountains.

In 1744 it joined the King's First Corps, marched to Bohemia and helped to take Prague on September 16, after a five-day siege. It remained in the capital as occupying troops under Lieutenant General von Einsiedel. As of October 4 the connection with the main army was interrupted. When the King had to draw back slowly, beginning on November 22, there was nothing the occupying troops could do but break out without wagons and artillery. Despite hunger, cold and other problems, the regiment fought its way by roundabout routes to the Silesian border, losing almost half its men. In the latter half of April 1745 it went with General von Hautcharmoy to Upper Silesia to rid the area north of the Oder of Hungarian raiders. On May 22 it took part in Margrave Carl's famous breakthrough from Jagërndorf to Neustadt, without fighting at Hohenfriedeberg. Upper Silesia again fell to the 'insurgents'.

In 1756 it remained at the old number of 1220 men; on July 25 it was mobilized in Far Pomerania under Lieutenant General the Hereditary Prince of Hesse-Darmstadt. to reinforce the Prussian Corps. But the Russians attacked only a year later. At Prague on May 6, 1717 six battalions, along with the grenadiers, were on the right wing of the second line under Hautcharmoy, forming the third reinforcement for the King in the breakthrough south of Kej, after Winterfeldt's first attack had been beaten back. With Prince Franz at its head, it followed I.R. 18 in storming the long causeway across the lowlands south of Hostawitz in the direction of Mount Tabor and came into the fighting at the Rokenitz Brook. At the end it advanced "to near the city" despite heavy losses, "to the King's greatest approval". The King wrote: "Every battalion must make a great step forward to harm the enemy. I am determined to remain strong in one position until something worthwhile can be decided". Kolin was to bring the decision: "Everything impelled me to advance most boldly to a conclusion". Here the grenadiers shed their blood, losing their commander. At the end of August I.R. 39, under Bevern, moved into Lusatia and Silesia. At the Lohe before Breslau one battalion fought on November 22. At Leuthen on December 5 it fought on the right wing of the second line, pushed ahead into the first line and captured fourteen cannon. In 1758, after the fall of Schweidnitz, it went with the King to besiege Olmtz. When it returned from Moravia it guarded the vehicle park against repeated attacks. At Hochkirch on October 14 it was with Retzow's Corps on the left flank, protecting the withdrawal. The grenadiers held the advance position at Kuppritz. Prince Friedrich Franz, its chief, was killed by a cannonball while on the right wing; he was a 26-year-old Major General. Despite outstanding achievements and the King's good will, the regiment saw no more service because of a lack of recruits; it was exhausted, though in 1759 it returned to the King's Army but reserved its power at Landeshut and Schmottseiffen. In 1760 it belonged to Prince Heinrich's Corps—again with two battalions—which gathered in May between Sagan and Lauban. The Battle of Liegnitz on August 15, where the grenadiers lost 389 men on the left wing, brought the King and his brother together. On September 17 it joined I.R. 35 in a "heated battle" at Hoch-Giersdorf, capturing seventeen cannon. With the King's Army in Saxony in the first months of 1761, it was surrounded by six times its strength at Altenburg, but was able to withdraw to Leipzig without losses. After the camp at Bunzelwitz—"a masterpiece of splendid field fortification"—the grenadiers took 5000 wagons and 1800 prisoners at Gostyn, while the regiment followedon the way to Pomerania only on November 14 and stormed the battlements at Spie on December 12. The march back via Stettin to Saxony exhausted the mauled regiment. From April 1762 on it was stationed in Neisse. In 1763 it had 561 Prussians, 211 Saxons and 1098 'foreigners'. The King showed it great favor in peacetime.

With Hohenlohe's Corps in 1806, it was disbanded in the surrender of Magdeburg on November 8, the Third Battalion at that of Glogau.

Füsilier-Regiment
Nr. 39
PRO GLORIA ET PATRIA
G. Dorr

INFANTRY REGIMENT NO. 40
Fusilier Regiment

Commanders of the Regiment

1740	10/17	Colonel Wilhelm Heinrich, Duke of Saxe-Eisenach
1741	8/31	Lieutenant General David Jürgen von Grevenitz, later Governor in Küstrin
1743	7/24	Major General Georg Christoph von Kreytzen, later Lieutenant General
1750	5/1	Colonel Johann Friedrich von Kreytzen, his brother, later Lieutenant General
1759	4/7	Major General Georg Carl Gottlieb von der Gabelentz, later Lieutenant General
1777	6/25	Major General Friedrich August, Baron von Erlach, later Lieutenant General
1791	5/20	Major General Carl Ludwig von Pfuhl
1794	12/29	Major General Friedrich Wilhelm von Steinwehr, later Lieutenant General
1805	10/19	Colonel Dietrich Leberecht von Schimonsky, later Major General

On October 17, 1740 Duke Wilhelm Heinrich of Saxe-Eisenach promised the King a regiment. The Duke was his uncle by marriage. The regiment, already founded in 1732 and having two musketeer battalions, had served the Emperor in the War of the Polish Succession, fighting on the Rhine, in northern Italy and at Lake Constance, and had had three battalions since 1734. After peace was made in 1738 and it had returned to Eisenach in 1739, it was reduced to one battalion, and most of its officers left. Promised to Prussia on October 8, 1739, it was taken command of by Colonel von Derschau in Magdeburg in 1740. The Duke engaged 25 'foreign' officers, provided replacements for the second battalion, and became chief of the regiment. After his death, his successor, Duke Ernst August II of Weimar, patron of Goethe and Schiller, declined the position, "since in sixteen years some fourteen regiments have been taken out of the duchies". Thus on August 31, 1741 it received Lieutenant General von Grevenitz. The King transferred the regiment to Berlin to discourage desertion, and in October he sent the First Battalion to Frankfurt, the Second Battalion to Küstrin and the grenadiers to Crossen, and kept an eye on them. In November it was at full strength. In February of 1742 the First Battalion was stationed in Glogau, the Second in Breslau and the grenadiers in Ohlau. But the regiment was soon concentrated in Breslau, staying there until 1756 and moving to Schweidnitz in 1763, with the grenadiers in Striegau and three companies in Cosel. Replacements came from the districts of Neumarkt, Goldberg and Haynau and their cities.

In the latter half of August 1744 it joined the King's First Corps to march through Saxony to Prague, which it besieged from September 11 to 16. Then it headed for southern Bohemia under General von Nassau, took Tabor, Weseli and Budweis, the last of which it occupied, plus Frauenberg Castle, where three companies served under Major von Conradi. When the King moved against the Saxons, these places could no longer be held. The evacuation order did not get through. After fighting off two attacks, Major General von Kreytzen had to surrender in Budweis with 930 men and many wounded and sick, as did Conradi with 284 men at Frauenberg. The regiment was imprisoned at Neuhaus. In 1745 the grenadiers threw the Saxons back at Hohenfriedeberg, attacking through the lowlands between Pilgramshain and Günthersdorf and to the right. At Soor they secured the Staudenz camp. Only in September was the regiment exchanged at Neubidschow and marched to join the army at Jaromer, then via Landeshut to Liegnitz. In the last months it guarded transportation to Görlitz and Breslau as well as Saxon prisoners from Kath. Hennersdorf. It was back in Breslau by the end of December.

On August 26, 1756 it marched into the Frankenstein camp to join the Second Corps of Field Marshal Count von Schwerin, which advanced to Jaromer in mid-September to restrain Piccolomini but returned at the end of October. On April 8, 1757 it marched to Bunzlau in General von Puttkamer's avant-garde, taking the magazine there on April 26. On May 6 it took part in the failed left-wing attack at Prague, which the King led to victory only after the bloody battle by the Rokenitz Brook. On May 17 the Second Battalion took a hill at Kolin "with great bravery". In the battle on June 18, in the second line on the right wing under Bevern, it got to Brzezan under heavy artillery fire and, almost at the end, was attacked by the Austrian cavalry, as was the First Guard Battalion. About 6:00 P.M. it returned to Planjan to keep the route to Nimburg open. It marched unstopped via Bömisch-Leipa, where the baggage train was lost, to Zittau under Winterfeldt, then to Schweidnitz under Grumbkow. Here it attacked and took 400 prisoners, but after the loss of several forts it was itself captured on November 12; its commander, Colonel von der Gabelentz, refused to sign the capitulation. While it was on its way to Brünn, the grenadiers were fighting on the right wing of the avant-garde at Leuthen. After its return in 1758 it was reorganized at Landeshut by July 1 and sent from the King's Army to Dohna's Corps at Frankfurt on July 31, moving to west of Küstrin on August 17.

At Zorndorf on August 25 the Second Battalion covered the exposed great battery on the right wing. The Russian cavalry attack led by General Demiku struck it on the flank; it was surrounded, losing two flags, and only saved by the counterattack under Seydlitz and Schorlemmer. Then the attack continued. The King was "highly satisfied". After that it was camped at Blumberg. While the grenadiers were captured while on the left wing near Lauske at Hochkirch, I.R. 40 marched to the Mulde, where it fought at Torgau and Eilenburg in November. On November 24 it moved into Near Pomerania, crossing the Recknitz at Damgarten on January 1, 1759 and taking Demmin and Anklam. In the Pomeranian Corps, it moved to Stargard in mid-May and took part in an advance to Thorn beginning on June 24, ending at Obornik. On July 23 it suffered high losses in repeated attacks on the Palzig Heights at Kay, so that at Kunersdorf on August 11-12 it guarded the Oder bridges at Görtz. At the end of September it was in Saxony again. Inm 1760 it besieged Dresden in vain and went off to Silesia on July 22, where on August 15 it served on the left wing at Liegnitz. The King wrote: "I shall end this campaign, determined to take any risk to find victory or an honorable end". In 1761 it made an advance toward Poland and was camped with the King at Bunzelwitz. Then it mopped up Upper Silesia. In 1762 it was still in action at the recapture of Schweidnitz from August 8 to October 10. In 1778-1779 Major General Baron von Erlach was its commander, a member of a Bernese patrician family who attracted unpleasant attention. In 1784 the King found that the regiment "did not resemble any soldiers"; in 1785 he put it more bluntly. In 1806 it was disbanded in the capitulations of Erfurt and Magdeburg, the Third Battalion in that of Schweidnitz. Men on garrison duty joined the Second West Prussian Regiment.

Füsilier - Regiment Nr. 40

INFANTRY REGIMENT NO. 41
Fusilier Regiment

Commanders of the Regiment

1741	6/8	Colonel August Wilhelm, Duke of Braunschweig-Bevern, later General of the Infantry
1741	10/31	Major General Johann Volprecht, Baron von Riedesel zu Eisenach, later Lieutenant General
1746	7/27	Colonel Franz Carl Ludwig, Count von Wied zu Neuwied, later Lieutenant General
1765	11/24	Colonel Matthias Ludwig von Lossau, later Lieutenant General
1782	5/20	Major General Alexander Friedrich von Woldeck, later Lieutenant General and Governor in Wesel
1792	11/7	Colonel Carl Friedrich von Schladen, later Lieutenant General
1804	9/24	Major General Carl Ernst von Lettow

After negotiations that began in August of 1740, the King took over the Württemberg "Hereditary Prince" Regiment, already in existence since 1716, on January 14, 1741. Originally Friedrich II had wanted only the rank and file without officers, not complete regiments. Taken over on March 1 and sworn in on May 2 at Lauffen on the Neckar, it went by ship to Wesel, arriving on May 12. Most of the officers came from the service of Württemberg. On June 8 Duke August Wilhelm of Braunschweig-Bevern became its chief, but on October 31 he passed it on to Major General the Baron von Riedesel, whom the King knew from the Imperial army that served on the upper Rhine in 1734 and was "very happy" to see as its chief. The negotiations had once again taken place among relatives. Ready on October 27, I.R. 41 gave two extra companies to Brieg for the foundation of I.R. 42. It stayed in Wesel until 1754, getting its replacements by recruitment. Since 1743 it, like I.R. 44 and 45, was no longer ranked with the field regiments, but was called the "Wesel Garrison" until 1763 and ranked ahead of all the garrison regiments. In 1755 it was transferred to Minden and given the recruiting area of Rahden, Reineberg, Hausberge and Petershagen with the cities of Minden and Lübbecke. As the King had recorded in 1768, the farmers of the Principality of Minden had registered themselves as soldiers and dealt with deserters by "casting them out with condemnation and compelling them to return to their regiments"—astounding for those times. In 1806 it was still stationed in Minden, with the grenadiers in Herford and three companies in Emden.

In Wesel the regiment had taken the place of I.R. 28, which had been transferred to Silesia. As of August 1, 1743 the King had the grenadier companies strengthened by 30 men, to be completed by April 1, 1744, and on May 20 of that year he assembled the grenadiers of I.R. 41, 44 and 45 from Wesel, and those of Garrison Battalions IX and XIII of Geldern and Minden, in Minden to be trained for field service. The three regiments in Wesel, though, were not not prepared for field service in 1743, not even with the Markish Corps. The formation of eight grenadier companies into a fusilier regiment in Minden in April of 1744, which the King had planned, fell through for financial reasons. Tactics were improved by the "Orders for the Generals of the Infantry" of July 23, in which was stated: "In a battalion there are two matters of concern: first, to form the line quickly, second, to keep the infantry advancing to gain more and more terrain on the enemy, for it does not depend on the dead buton the place". At the end of April 1745 the Minden Grenadiers, commanded by the Old Dessauer, were at Magdeburg ready to fight the Saxons and constantly strengthened by the King. In the attack on Kesselsdorf on December 15 the Münchow Grenadier Battalion alone lost 197 dead and 179 wounded. In 1747 it returned to Wesel via Magdeburg and Aken; as of 1748 it remained in Magdeburg permanently, always under the eye of I.R. 3.

According to the King's measures of June 21, 1756, I.R. 41 alone was sent to the army from Wesel, and with heightened fighting strength. On the right wing under Duke Ferdinand of Braunschweig, it marched through Saxony via Nollendorf to Aussia to take part in the siege of Pirna. Strengthened by 300 men as of January 9, 1757, it went to Bohemia in mid-April and was part of the second line on the right wing at Prague on May 6, providing the last reinforcement for Hautcharmoy's Division in its breakthrough south of Kej, which led to the infantry battle at Hrdlorzez. At Kolin on June 18 with the grenadiers, it was taken out of the second line on the left wing by the King and sent into the gap between Prince Franz's and Pannewitz's Brigades for a new attack on the Krzeczhorz Heights. Led by the King with drawn sword, and with artillery support, the bitter storming succeeded about 5:30 P.M. In the tragic finale at Novemesto, the grenadiers secured the retreat of the second line's right wing with the artillery, wounded and fugitives. The regiment had lost 1018 men, including Colonel Herwarth von Bittenfeld, its commander. In mid-August it went with Bevern to Silesia. On September 7 the grenadiers defended the Jäckelsberg at Moys, losing 204 dead and 68 wounded. On November 22 the regiment stood its ground at the Pilsnitz Forest near Breslau, and at Leuthen it fought steadfastly on the right wing between the two lines, losing 79 dead and 73 wounded. In 1758 the entire regiment besieged Schweidnitz, whose Gallows Fort the grenadiers stormed on April 16. Then it vainly besieged Olmütz from June 1 to July 1, while the grenadiers fought at Domstadtl. On the march back through Königgrätz to Landeshut it captured Kronau, which was occupied by eight grenadier companies, and opened the way. On August 11 ther King took it along to join Dohna's Corps. While he crossed the Oder to circle around to the east of the Russians, it occupied the shattered city of Küstrin and then, under Prince Franz at Beeskow, secured Lusatia against the Austrians, going on to Lübben on September 4. At Hochkirch on October 14 the grenadiers on the left wing were almost wiped out in the first attack, while the regiment, under Wedell, drove back the Swedes at Fehrbellin from October 22 to 28, then moved via Prenzlau to arrive at Torgau on November 12 and secure the Mulde. In 1759 it went to Franconia with the Saxon Corps but then went along to Kunersdorf, where on August 12 it suffered very heavy losses in the Kuhgrund, including Colonel von Böhm, its regimental commander. At Maxen the grenadiers were taken prisoner. At Liegnitz on August 15 it led the counterattack on Panten; its commander received the Order of the Black Eagle. At Torgau on November 3, under Zieten in the Tettenborn Brigade, it took the village of Süptitz and attacked the enemy from the south despite considerable losses. In 1761 and 1762 the grenadiers fought in Pomerania and Upper Silesia, while the regiment fought under the King, lastly on August 16 at Peilau.

In 1763 I.R. 41 consisted of 1093 Prussians, 125 Saxons and 402 'foreigners'. The King showed them all favor. After 25 years in the service of Wrttemberg and 65 in that of Prussia, it was disbanded in 1806 in the capitulations of Hamelin on November 22 and Nienburg on November 26, including the Third Battalion.

Infanterie-Regiment
Nr. 41

INFANTRY REGIMENT NO. 42
Fusilier Regiment

Commanders of the Regiment

1741	11/3	Major General Heinrich, Margrave of Brandenburg-Schwedt, Chief until 12/12/1788 Commanders:
1741	11/3	Lieutenant Colonel Martin Anton, Baron von Puttkamer, later Major General
1750	9/24	Colonel Johann Christian von Conrady
1753	5/28	Colonel Balthasar Rudolf von Schenckendorff, later Major General
1760	4/22	Colonel Heinrich Werner von Kleist, later Major General
1764	4/2	Major Heinrich Wilhelm von Lettow, transferred to I.R. 46
1776	10/20	Lieutenant Colonel Stephan von Lichnowski, transferred to I.R. 23
1786	3/1	Lieutenant Colonel Georg Bogislav von Kthen, transferred to I.R. 48
1789	2/7	Colonel Nikolaus Reinhold von Hahnenfeldt, later Lieutenant General
1794	12/29	Major General Caspar Siegmund von Lüttwitz
1796	4/5	Colonel Christian Franz Heinrich von Plötz, later Lieutenant General

The Garrison Regiment No. 6, taken command of by Colonel von Stechow on July 17, 1741, had been formed in Brieg on May 4, after the Battle of Mollwitz and the taking of the city, and was composed of 400 men from the occupying forces—some volunteers, some not. They had been granted free departure with the condition that they not fight against Prussia for two years. Colonel von Wurmb directed the formation. When Stechow took over the regiment, eight hundred "whitecoats" (laboring soldiers) of the Mützschefahl Detachment who had been taken prisoner by the Austrians at Grottkow before the battle and just exchanged on July 9, had been assigned to it. The grenadiers came from the "Hereditary Prince" Regiment of Wrttemberg, now I.R. 41. Margrave Heinrich of Brandenburg-Schweydt, whom the King had reassigned from I.R. 12 on May 31, 1741, was named to be its chief. The manpower was anything but elite. Margrave Heinrich (1709-1788) was, of course, a Major General and was given the regiment on November 3 of that year, but had lost the esteem of the King because he frequently reported sick. He drew his salary, but was not actually allowed to command the regiment. For that reason, the regiment was the only one in the army not to have a guard company. The Margrave was the son-in-law of the Old Dessauer and a cousin of Friedrich Wilhelm I and of Margrave Carl, whom the King particularly esteemed. The garrison regiment was filled aby the end of 1742 with recruits from Upper Silesia and Bohemia. After peace was made on June 11, 1742, I.R. 42 was transferred to Jauer and the surrounding area. From 1746 to 1756 it was in Reichenbach and Frankenstein, from 1763 to 1781 in Frankenstein, then in Breslau until 1787 and Neisse until 1795. Since January 6, 1796 the regiment was in Warsaw with three companies in Tschenstochau. Its replacements came from the Upper Silesian districts of Beuthen, Gross-Strehlitz and Pless with their cities until 1797, then from the districts of Gostyn, Rawa, Sochaczew and Czersk plus parts of Warsaw and Blonie with their cities.

As of August 4, 1742 each regiment had its own fixed canton in Silesia. On January 11, 1743 this ruling was renewed, declaring "all actually domiciled burghers and farmers free from the draft (in peacetime) without any exception". General von der Marwitz, Commanding General in Silesia from March 27, 1743 until succeeded by General von Buddenbrock in 1745, enforced the rules honestly. But the King kept Silesia under his own control and visited it annually, so that the generals there were not very independent. In 1744 I.R. 42 joined von der Marwitz's Second Corps, which gathered between Neisse and Neustadt to secure Upper Silesia but did not advance to Olmütz. The grenadiers were with the regiment. In 1745 it fought under Nassau, then Margrave Carl, who returned to Neisse in April. The First Battalion lost 24 deserters in the process, the grenadiers 76. At the beginning of May it and other regiments were assigned to occupation duty, presumably of the Fortress of Schweidnitz.

With Field Marshal von Schwerin's Second Corps in 1756, it marched through the Silesian mountain passes to the Elbe and Adler in mid-September but saw no action, left at the end of October and moved into winter quarters in early December. On January 9, 1757 it was strengthened by 300 men. Schwerin crossed the mountains between Schmiedeberg and Glatz and went to meet the King at Prague. On May 6 the regiment fought without particular distinction in the second line on the left wing, which the King led toward Prague on a broad front along the Rokenitz Brook after the battle. The grenadiers secured the Isar bridges at Zamost and went down fighting at Kolin on June 18: one officer, four non-commissioned officers and nineteen men were unharmed, thirty-nine wounded were saved from being captured. It was refilled with Saxon soldiers. On the march back under Prince August Wilhelm, the Second Battalion, with Colonel von Diericke, was taken prisoner on July 22 in the ruins of Zittau. In 1758 I.R. 42 belonged to the King's Army after the prisoners had returned via Jägerndorf in mid-April. In Keith's Corps it encircled Olmütz from May 20 to July 1 from the west, until the King withdrew because of changes in the military situation. At the beginning of August it went back to Silesia with him via Leitmeritz, Königgrätz and Landeshut, and stayed there under Margrave Carl from August 10 on. In 1759, under Major General von Wobersnow, it led the advance from Glogau to Posen to destroy the great Russian magazine and fought at Konradswalde southwest of Landeck in the County of Glatz. From July 10 to August 16 it was in camp at Schmottseiffen, under the command of Prince Heinrich as of July 29. The King's willing courtesy affected the soldiers' spirits. "The old warriors' faces brightened up when he wished them a good morning after an exhausting march. In time of great danger he remained cheerful, quiet, calm, his eye always on the officers and soldiers."

In 1760 the First Battalion shed blood under Fouqué at Landeshut; the survivors were taken prisoner. The Second Battalion was with Zieten in the Freiburg area. The grenadiers fought at Dresden, Liegnitz, Hochgiersdorf and Torgau. Their commander received the Pour-le-merite and became a colonel. In Goltz' Corps in 1761, still "on the old foot", it left on September 12, led by Platen, via Gostyn to Kolberg and took part in the Pomeranian campaign. In 1762 the King delayed the start of the campaign. His alliance with the Russians on June 30 made caution unnecessary. On August 16, under Bevern, it successfully dealt with four Austrian corps north of Peilau. The Second Battalion, sent to Girlsberg under Captain von Czettritz, fought to the last man on the left wing. When the King arrived, the enemy departed.

In the L'Estoq Corps in 1806-1807, its numbers sank on account of the desertion of Polish soldiers after Tilsit; what remained joined the First Silesian Infantry Regiment on November 21, 1808, in 1914 the First Grenadier Regiment King Friedrich Wilhelm II (First Silesian) No. 10 of Schweidnitz.

Infanterie-Regiment
Nr. 42
Füsiliere

INFANTRY REGIMENT NO. 43
Fusilier Regiment

Commanders of the Regiment

1741	8/10	Major General Maximilian von Rampusch
1743	2/15	Colonel Philipp Ernst von Bardeleben
1744	7/2	Colonel Joachim Ernst von Zimmernow, died 10/22/1744 after Pardubitz
1744	12/9	Colonel Christian Ludwig von Kalsow, later Lieutenant General
1757	1/19	Major General Samuel Adolph von Kalckreuth
1758	1/22	Colonel Joachim Leopold von Bredow, later Major General
1760	2/5	Major General Christian Wilhelm von Zieten
1767	6/17	Colonel Döring Wilhelm von Krockow, later Major General
1773	1/11	Colonel Carl Magnus von Schwerin, later Major General
1776	1/24	Colonel Leopold Ludwig, Count of Anhalt, later General of the Infantry
1795	5/19	Major General Alexander Leopold, Count von Wartensleben, later Lieutenant General
1803	1/24	Major General Ludwig von Strachwitz

Ruled by Bohemian kings in the Middle Ages and by the Hapsburgs since 1526, Breslau had developed into a city-republic with its own administration. A ducal residence since 1163, a Hanseatic city, a principality of the Empire since 1280, with its bishop elevated to Prince of Neisse-Ottmachau in 1290, under foreign rule since 1327, the city had been spared any siege or ruler's garrison. For that reason she kept her city soldiers ready but outside the city, because the citizens provided their own watchmen and defenders. After years of construction its fortress was finished in 1576, excluding the Cathedral Island. On December 14, 1740 an uprising prevented the entry of three Austrian battalions and any defensive measures. The council even had all gunpowder shipped to Brieg on December 29. On January 1, 1741 the King was before the city, occupying Cathedral Island while negotiations went on. On January 3 he signed the "Treaty of Neutrality": no oath of allegiance, no admission of Austrians, admission of individual Prussian soldiers only "without upper guns", passage through the city of only one company at a time, escorted by the city militia, establishment of a main magazine in the suburbs with one battalion to guard it, formal entry of the King with thirty gendarmes as a 'guest'. Austria was surprised. "Intrigues" changed the situation. At one stroke the inner city was occupied on August 10, the burgher companies were disarmed and the 600 city soldiers were sworn in as the cadre of the new Garrison Regiment 5 of Breslau and Glogau, with Colonel von Bardeleben as Commander and Major General von Rampusch, Commandant of the City, as its chief. One stroke of the pen! On July 21, 1743 the King held his first "Silesian General Review" at Hundsfeld, northeast of Breslau. On August 1, 1744, a month after Colonel von Zimmernow took command, it became Fusilier Regiment 43. From 1745 to 1756 it was in Schweidnitz, from 1763 on in Liegnitz, with the grenadiers in Striegau, later with three companies in Nimptsch. Its recruits came from the districts of Wohlau, Steinau and Militsch and their cities. On August 16, 1744, two weeks after being changed to a field troop, it marched with Schwerin's Corps via Braunau and Königgrätz, although intended for a communications zone, to Prague, where the grenadiers took part in the siege as of September 16. During an unprepared withdrawal, the outnumbered First Battalion bravely fought off attacks on the great magazine at Pardubitz on October 19; Colonel von Zimmernow died of his wounds on the 22nd. The grenadiers fought their way through with only 279 men. In 1745 I.R. 43 secured the fortress and magazine at Schweidnitz, between Breslau and Bohemia. At Crossen on August 14 the grenadiers fought off Saxon cavalry, guarded the withdrawal across the Oder to Glogau, and appropriated the Saxon magazine at Guben on November 29.

Training in peacetime stressed marching and advancing at a steady pace: "While shooting, the people must always be driven forward, thus forcing the enemy to retreat, whereby confusion inevitably follows among them". In the Infantry Regulations of June 1, 1750 it was stated in Article V: "When fellows are invalided because of age or illness, then His Majesty does not want them to be discharged without their advance knowledge; rather the regiments should report it, so that they can be cared for and not have to go begging. N.B. . . . discharged soldiers should keep their coats, shirts and trousers, so the soldier is not compelled to go naked from the regiment".

In Schwerin's Silesian Corps in 1756, its grenadiers ended enemy attempts to pin them down, which had begun in mid-September, on October 29 with the rear-guard action at Lewin. On January 9, 1757 it was strengthened by 300 men, and in mid-April it marched along to Prague. Grenadiers and regiment, on the left wing, were involved in an impromptu attack and beaten back, losing 400 men, and were then used to reinforce the siege of the city from May 6 to June 20. On the way home the First Battalion, barely 450 men strong, had reached Gabel, which defended the link with Zittau, when they were attacked for 24 hours by vastly superior forces and taken prisoner on July 15. At the end of August the grenadiers and Second Battalion, under Bevern, marched to Silesia. Before Breslau on November 22 they bravely maintained their position on the right wing at Pilsnitz under General von Brandes. After the capitulation of Breslau the Second Battalion had only a hundred men left. Back to full strength and in the Saxon Corps in 1758, it took part in the invasion of Franconia as far as Hof and, from October 12 to 14, Wedell's counterattack on the Mulde crossings at Torgau and Eilenburg. The grenadiers took Schweidnitz on April 16 and fought at Zorndorf on August 25, where it lost 246 men and 87 more were captured. In 1759 it was in the avant-garde of the corps at Himmelskron, near Kulmbach, opposing the Imperial army until May 11. At the end of July it went through Bautzen with the King to oppose the Russians on the Oder. At Kunersdorf on August 12, in the second line of the avant-garde before the right wing, it overran the Bäckergrund with its abatis after a half-hour of artillery fire and stormed the Mühlberg, with its great battery of 600 guns, despite withering fire. The attack came to a stop in the Kuhgrund, shattered by enemy attacks, with 550 men lost, including Major the Prince of Holstein-Beck. It had "come out very advantageously" for the King. In 1760 it was with Prince Heinrich opposing the Russians between Landsberg and northern Silesia, in 1761 in camp with the King at Bunzelwitz. In 1762 I.R. 43 captured the 460-meter-high Eisenberg at Adelsbach, southwest of Freiburg, despite heavy fire and the loss of 400 men. Although the King broke off the battle, his verve forced Daun to retreat. The King excluded the regiment from his 1784 criticism of the Silesian regiments. On August 27, 1785 he appointed its chief, Lieutenant General the Count of Anhalt, Inspector of the Lower Silesian regiments, including his own. In Rüchel's Corps in 1806, it was disbanded in the capitulations of Magdeburg, the Third Battalion by that of Schweidnitz. The men on garrison duty joined the Second West Prussian Regiment.

Füsilier-Regiment
Nr. 43
PRO GLORIA ET PATRIA

INFANTRY REGIMENT NO. 44
Fusilier Regiment

Commanders of the Regiment

1742	1/30	Major General Friedrich Ludwig, Count zu Dohna, later Field Marshal
1749	1/14	Colonel Martin Eberhard von Jungkenn Müntzer von Mohrenstamm, later Major General
1759	1/4	Colonel Rudolf August von Hoffmann
1760	2/5	Major General Johann von Grant, later Commandant of Neisse
1764	12/9	Colonel Hans Christian von Brietzke, later Lieutenant General
1779	6/17	Colonel Friedrich Wilhelm Ernst von Gaudi, later Lieutenant General and Commandant of Wesel
1789	2/7	Colonel Franz Otto von Pirch, transferred to I.R. 8 as a Major General
1791	8/31	Colonel August, Count zu Dohna, later Major General
1793	1/9	Major General Franz Georg von Kunitzky
1799	2/14	Colonel Ludwig von Strachwitz, transferred to I.R. 43
1803	5/3	Colonel Christian Alexander von Hagken, later Major General

After Colonel von Kalnein had done a successful job of recruiting for I.R. 41, he was given the task of raising a second regiment in Württemberg, to whose reigning family that of Prussia was related. The regiment was intended to replace I.R. 28 in Wesel. Because of the high costs, Kalnein raised only a small number of recruits; the majority of men were recruited elsewhere in the Empire. On December 28, 1741 the King ordered Major General Friedrich Ludwig, Count zu Dohna, to send I.R. 28 marching to Silesia. He was allowed to keep ten men per company to form the core of the new regiment, and was granted the old regiment's uniforms and recruiting area. On January 30, 1742 Count Dohna became its chief; in July of 1742 it was filled and given the number 44. In the same year the King made further negotiations with the lesser German princes to obtain soldiers. Because of its proximity to the heavily populated areas of Cologne, Düsseldorf and Westphalia, Wesel was a good place to establish a regiment, even though the men recruited from that triangular area were not of the best quality. On the other hand, enlistment rules and cantonal policies in Cleves, Geldern and Mrs on the subjects of occupation, religion and areas of Prussian possession could hardly be changed. Prussia was officially a Protestant state, with full tolerance. Thus in 1747 the expected replacements had to be done without; the regiment remained on the "old foot" with 1220 men and without a recruiting area. From 1743 to 1763 it was excluded from field duty as the "Wesel Garrison". The grenadiers were stationed in Magdeburg since 1746. Standards were scarcely higher than in garrison regiments. It was not called on in 1744 or 1756. In 1749 the King turned the regiment over to Colonel von Jungkenn Müntzer von Mohrenstamm, who had entered the service of Württemberg from that of the Palatinate in 1713, had been with I.R. 41 since 1716 and taken part in the campaigns, had been accepted as a Lieutenant Colonel in 1741 and commanded I.R. 41. Two of his sons served in his regiment, two in others. Wesel remained the garrison site until 1805, when it was moved to Münster and the grenadiers went to Soest and three companies to Drensteinfurt and Ahlen. The Third Battalion, transferred to Warsaw in 1800, formed the basis of I.R. 60 in 1802. In exchange, the regiment in Mnster was given a new third battalion.

Though the regiment was unaffected by the two Silesian Wars, the grenadiers left Magdeburg on August 21, 1745 to join Prince Leopold of Anhalt-Dessau's Corps at Gattersleben. After taking Torgau on December 5, they fought at Kesselsdorf on December 15 in the first encounter of the right flank. Their attack on the ice-covered heights broke down under fire; Battalion 41/44 lost 197 dead and 179 injured, and had just a hundred men left. The King had said: "Of all the means that I have at my disposal, war is the only suitable one. This method will bring a decision within a few hours". Seven years of war still awaited him. The grenadiers were stationed in small towns around Magdeburg, then in Magdeburg itself as of January 1748, with its own budget as well as support from the regiment.

Not called on in 1756, I.R. 44 fought in secondary actions in 1757: while the grenadiers fought at Prague, Breslau and Leuthen, the regiment, once it had left Wesel, joined the Allies Corps, commanded by the Duke of Cumberland, at Bielefeld and moved ahead of the Imperial Army. In a night rear-guard action it lost 200 men, and 220 deserted as it returned from Minden to Magdeburg. Replacements from Halberstadt and Magdeburg brought its numbers back up to 1320 men. After Rossbach, Duke Ferdinand of Braunschweig took command of the corps, which had been disorganized by the Zeven Convention of September 9. In 1758 it followed the Saxon Corps "for purposes of occupation" and fought the Swedes at Fehrbellin under Wedell. By September 28 the Rhin sector was reached; as of November 12 it was at Torgau again. In the Saxon Corps with the same job to do in 1759, the Second Battalion, which had occupied Breslau, evacuated the city in the "free departure" in order to hold Torgau and its magazine. After a brave defense againbst the Imperial army, it surrendered there on August 15 and moved away. The First Battalion secured Dresden, which General von Schmettau handed over on September 4, although the King was on his way. When his chief, Colonel von Hoffmann, who went on guard duty at the castle drunk, called out to it: "We march out like scoundrels and you are really lazy" and fired his pistol without hitting anyone, the guards shot him from his horse. The King passed judgment: "I am of Hoffmann's opinion; he could not have expressed himself otherwise about the evacuation!" As of September 26 it served with Finck's Corps and fought at Korbitz and Strehla. In 1760 I.R. 44 belonged to the King's Army and took part in the unsuccessful siege of Dresden from July 10 to 22. On August 20 it fought off the Imperial army's attacks on the heights at Strehla, under Lieutenant General von Hülsen, and defended Torgau until September 26. Hülsen reported: Our troops, from officers to common men, did what honorable people are supposed to do", thorough praise. In 1761 it fought in the Saxon Corps and experienced the capable leadership of Prince Heinrich against the Imperial and main Austrian armies along the Mulde. In 1762, after a short offensive, it took part in the attack on Freiberg on May 12 and on the heights of Gr. Schirma on October 29, where it was in von Taube's Brigade at the axis of the army's swing to the left in the regiment's only major battle.

In 1806 it was disbanded in the capitulation of Le Coq's Corps on November 22 in Hamelin, including the Third Battalion. In Cosel and Neisse on November 21, 1808 its remaining men joined the First Silesian Infantry Regiment, later the Grenadier Regiment King Friedrich Wilhelm II (First Silesian) No. 10 of Schweidnitz.

Füsilier-Regiment
Nr. 44

INFANTRY REGIMENT NO. 45
Fusilier Regiment

Commanders of the Regiment

1743	8/1	Lieutenant General Friedrich Wilhelm von Dossow, later Field Marshal
1757	1/8	Lieutenant General Hereditary Prince, since 2/1/1760 Landgrave Friedrich II of Hesse-Kassel, later Field Marshal
1786	3/1	Major General Johann Ludwig von Eckartsberg
1792	4/10	Colonel Friedrich August von Grevenitz, later Major General
1795	7/6	Colonel Carl Philipp von Unruh, later Lieutenant General
1805	10/19	Colonel Johann Christian von Zweiffel, later Major General

On January 23, 1743 Lieutenant General Friedrich Wilhelm von Dossow received the order to send his Fusilier Regiment No. 31, later musketeers, from Wesel to Silesia and to set up a new regiment in their place on August 1. Ten men from each company stayed behind to form the cadre of it. I.R. 31's uniform and recruiting area were turned over to the new regiment, which became No. 45. Some of the officers came from the army, others from foreign service. Men were recruited. All the special features of the "Wesel Garrison" applied here as to I.R. 44. As of May 1744 the grenadiers were transferred to Minden, formed a separate troop and moved to Magdeburg later. In 1747 it lost its canton and was obligated to recruit. It is noticeable that the King entrusted forming new regiments and recruiting for them to the same officers again and again. In 1743 it was not in field service, in 1744 it remained inland at Wesel, and in 1745 it was not called to the Old Dessauer's Corps either. Giving its command to the Hereditary Prince of Hesse-Kassel in 1757, who became Landgrave Friedrich II in 1760, was done for political reasons, though he had turned Catholic. He was to be kept in the Prussian-English alliance by satisfying his ambition with the rank of Lieutenant General and the governorship of Magdeburg, one of the most important fortresses, though he was freed from all duties. A field regiment was too high a price for the King to pay. Commanding I.R. 13 under Tsar Peter III in 1762 under similar but politically different circumstances was good enough for him. In 1756 the King had already granted the Hereditary Prince the new Fusilier Regiment No. 48—of the Salmuth Garrison Battalion in Minden—and then, when he was dissatisfied, let him have I.R. 45 after Field Marshal von Dossow had left it. The most remarkable favor was the retention of oboes when they were removed from all the other Wesel regiments. The careful treatment of relatives of the Swedish crown went a long way. When Ansbach-Bayreuth became Prussian in 1792, the regiment moved to Bayreuth, the Third Battalion to Hof, and the grenadiers to Kulmbach. Soldiers from Cleves and Mark remained in Wesel.

In 1756 the regiment stayed in Wesel on the "old foot" of 1220 men. Prussian superiority also resulted from better organization: "Because all colleagues, all departments, every subordinate pulled the same string, the King's organization was as good as infallible and always prompt. Thus he was always ready sooner than the Austrians and even more so than their allies, the Swedes and Russians". When the French army marched from the lower Rhine in April, Wesel was evacuated on March 24, 1757. Colonel von Salmuth defended well-protected Geldern with all his strength from April 1 to August 23. The "Prussian Brigade", including I.R. 45, left the Duke of Cumberland's Corpswhen he crossed the Weser. It avoided defeat at Hastenbeck and went to Magdeburg, where its loss of 320 deserters was replenished and the "middle foot" of 100 additional men was attained. In 1758 it advanced into Franconia and the Upper Palatinate with the Saxon Corps and withdrew to a position at Zschopau, then to Pirna on the Müglitz, where it stayed until September. Under General von Wedell it took Fehrbellin on September 28 and reached Torgau via Templin and Berlin on November 12, shortly before the Imperial army, to be reinforced by Dohna as of November 14. Torgau was an important site. On May 23, 1759 I.R. 45, under General von Schenkendorff, drove the Austrian Brentano Division back over the Erzgebirge at Aue, southeast of Zwickau. Early in August, under Colonel von Wolffersdorff, it staunchly defended Torgau against Prince Carl zu Stolberg despite numerous losses, and was granted an honorable capitulation with free departure: "When at the evacuation on August 15 his followers tried to induce Saxons and deserters of the Imperial army to leave it, Wolffersdorff shot some of them down and threatened to take the Prince prisoner". Thereupon he could continue the evacuation. Under General von Wunsch, who summoned the occupying forces from Wittenberg and Torgau, it retook Wittenberg and Torgau in night attacks but arrived too late at Reichenberg to prevent the surrender of Dresden. Wunsch returned to Torgau instantly and drove away the Imperial army of 14,000 men at Zinna with 5000 men by forming what appeared to be a broad front. Half of the 600 prisoners entered Prussian service at once. When Finck took over the corps, he had Wunsch occupy Leipzig and drove back superior enemy forces at Korbitz on September 21. Under Wunsch at Pretzsch on October 29 it fell to the enemy avant-garde; 1300 men were taken prisoner. These battles kept the King's rear free for the major battles. On November 18 Finck's Corps reached the flatlands at Maxen on the Müglitz, a daring move on the King's part. Attacked on all sides on November 20, Finck decided to surrender on November 21 after heavy losses, brave defense and many desertions; the regiment was included, though some men fought their way out.

In 1760 one battalion was with Stutterheim's Corps on the Ucker, then in Berlin opposing the Russians from October 5 to 9, at Kemberg on October 28, where the King realigned the army, and from October 31 on it occupied Leipzig. Regiments without canton areas back home were at a disadvantage. In 1761 it moved into the fortified camp at Kolberg with the Württemberg Corps at the beginning of June, but it could not be held despite reinforcements. Enduring hunger and cold, it broke free on November 14 and reached Treptow, to make the last attempt to relieve Spie on December 12 and then withdraw to Stettin and Lusatia. In 1762 the King began to fight only on July 1, with the help of the Russian supply corps. On August 16 its two battalions fought in Bevern's Corps north of the Peile, between Fischerberg and Spitzberg, beating back at attack of heavily superior forces. Finally Schweidnitz fell. In 1763 it had 76 Prussians, 95 Saxons and 779 'foreigners', a total of 950 men, on field duty. In 1806 it was disbanded in the capitulations of Erfurt and Magdeburg, the Third Battalion in that of Erfurt.

Infanterie-Regiment Nr. 45

INFANTRY REGIMENT NO. 46
Fusilier Regiment

Commanders of the Regiment

1743	4/8	Colonel Carl Eugene, Hereditary Prince of Württemberg, later Lieutenant General and reigning Duke (chief until 1757) Commanders: Colonel Ernst Ludwig von Götze, later Major General, died at Kesselsdorf
1745	12/26	Colonel Friedrich Leopold von Schwerin, later Major General
1750	6/14	Colonel Christian Ludwig von Pfuel, later Major General
1756	11/18	Major General Emanuel von Schöning, died 5/16/1757 of wounds at Prague
1757	May	Colonel Johann Albrecht von Bülow, chief since 12/26, later Lieutenant General
1776	10/20	Colonel Heinrich Wilhelm von Lettow, later Major General
1779	6/17	Colonel Ernst Ludwig von Pfuhl, later Lieutenant General and Governor of Spandau
1794	12/29	Major General Alexander Heinrich von Thile, later Lieutenant General

On February 28, 1743 the King—as in 1740—gave the regiments of the central provinces the task of recruiting for a new regiment from August to October, for delivery in Berlin, at a rate of ten Taler per man, exclusively 'foreigners'. On April 8 Hereditary Prince Carl Eugene of Württemberg-Stuttgart (1728-1793), reigning Duke as of February 1744, residing at the Prussian court from 1741 to 1744 and married to a niece of the King, and allied with France since 1752, became its chief. He also allowed recruiting in Württemberg. Because of his youth he did not actually command and left the chief's work to the commanders. When in 1757 he joined the side of Prussia's enemies, the King relieved him of his position and the regiment of its name. Formation of it took place in Nauen and Brandenburg. Transferred to Berlin in May of 1744, it was given the canton of I.R. 37, which had been sent to Silesia: the area around Brandenburg/Havel, but without the city. In Berlin it had been garrisoned in Neukölln and Köpenick, an advantage for the young regiment. Good recruitment came from the district of Ziesar, parts of the Zauche and Havelland districts and the cities of Werder and Ziesar, and from 1773 on also parts of the districts of Kamin, Deutsch-Krone and Hohensalza with their cities. In January of 1796 it was transferred to Warsaw, with three companies in Prague. Now the recruits came from the districts of Lencziz, Zgierz, Orlow, Brzezin and Petrikau and their cities, plus Gostynin and in part Warsaw, including many ethnic Germans. The officers did not have an easy job of creating a functioning troop with a common spirit out of such a variety of recruits. The King expected honor, loyalty and active combat readiness of them, and welcomed many-faceted education, depth of knowledge and intelligence. He put particular emphasis on the officers' orderly organization, being well aware of the relationship between a disorderly life style and careless service and treatment of subordinates.

At the beginning of the war in 1744 the "Alt-Württemberg" Regiment stayed in Berlin with the Guards and three other regiments. On September 25 it was placed under Prince Leopold of Anhalt-Dessau, who put a corps together at Magdeburg late in April of 1745 to use against Saxony. Ready to march as of July 23, it stayed at Gattersleben and Dieskau until November 23. The corps, on whose striking power the King based his hopes for peace, was steadily enlarged. On the morning of December 15 I.R. 46 stormed uphill over the Kessel-Bach at Kesselsdorf on the left flank, losing 600 men. Mobilized under the Hereditary Prince of Hesse-Darmstadt as of June 25, 1756 as a reserve for the Prussian Corps in Far Pomerania, it moved into Lusatia on December 7. The Russians arrived only on August 1, 1757. It was not enlarged on January 9, 1757, but remained on the "middle foot" with 1320 men. At Prague on May 6 the First Battalion fought on the left flank, which won the battle under the King's leadership, the Second was among the troops that surrounded the city; both suffered heavy losses. On the way back, the Second Battalion and the grenadiers were captured on July 15 after a brave defense at Gabel. In mid-August the First Battalion marched to Silesia with the Duke of Bevern. Passing Moys and Barschdorf, it reached a position on the Lohe near Breslau at the end of September. After an undeserved setback on November 22, it returned with Zieten to join the King at Parchwitz. "Not a one of them grumbled, rather they all withstood all difficulties with joy and good will". At Leuthen on December 5 it fought in the second encounter of the right wing under Major General von Bülow, its future chief. Its first few battles had hit it hard.

In 1758 the prisoners returned, and from late March to April 16 the regiment took part in the siege of Schweidnitz. From the beginning of May on it besieged Olmütz with the grenadiers, which ended in a march back to Silesia after Domstadtl. On July 31 it reached Dohna's camp at Frankfurt, while the grenadiers arrived with the King on August 11. At Zorndorf on August 25 it was hit from the side while on the right wing in the Graugreben Brigade. When the whole wing fell back, the King tried to inspire them to stand their ground with a regimental flag and his drawn sword. Seydlitz's counterattack gave them a respite. The regiment "gained the highest satisfaction". On September 2 it marched away via Küstrin and was just short of Dresden on September 11, covering 225 kilometers in eight days. "That the Prussians fought with the greatest bravery is proved by a mass of small incidents, but especially the astonishing, lasting, almost completely exhausting marches", Tempelhoff judged. On September 28, under Wedell, it took Fehrbellin from the Swedes to protect Berlin and then reached Torgau by November 12, before the Imperial army, so that the Mulde was held. In 1759 it was in Prince Heinrich's Saxon Corps that advanced into Franconia until early June. Then it was kept at Bautzen in readiness for the march to Kunserdorf on July 31. In the battle of August 12 it suffered heavy losses before the Great Spitzberg in the face of Russian steadfastness, great heat and shortage of ammunition. For the rest of the year it was again between Strehla and Torgau. On June 23, 1760 it went down to defeat under Fouqué at Landeshut "with the greatest bravery" against three times its numbers. Only a few men fought their way out, and no prisoners were exchanged. In 1761 and 1762 what remained of the battalion was stationed in Breslau, and was called on to take part in the siege of Schweidnitz. In 1763 it was filled with the remaining men of the Heer Free Battalion, Swiss deserters of the French army.

It was disbanded in the capitulation of Breslau on January 5, 1807 after defending the city under its last commander since November 8, 1806. The Third Battalion joined the First Silesian Infantry Regiment in East Prussia.

Infanterie-Regiment
Nr. 46
Füsiliere

INFANTRY REGIMENT NO. 47
Fusilier Regiment

Commanders of the Regiment

1743	8/1	Colonel Georg Wilhelm, Prince of Hesse-Darmstadt, later Major General
1747	2/26	Colonel Carl Friedrich von Derschau, later Major General
1752	8/16	Major General Leopold Friedrich Ludwig von Wietersheim, transferred to the Saxon Rochow Regiment in 1756
1756	10/20	Colonel Caspar Friedrich von Rohr, died 12/12/1757 of wounds at Leuthen as Major General
1758	1/5	Major General Christoph Heinrich von Grabow
1764	1/6	Major General Johann Adolph, Prince of Nassau-Usingen, later Lieutenant General
1778	11/23	Major General Wenzeslaus Christoph von Lehwaldt, later Lieutenant General
1788	6/1	Major General Friedrich Wilhelm von Wangenheim
1790	9/26	Major General Johann Carl, Count von Hertzberg
1797	9/12	Colonel Julius August von Grawert, later Lieutenant General

The second regiment for which the King ordered recruitment as of August 1, 1743 was granted by the order of August 1 to Prince Georg Wilhelm of Hesse-Darmstadt. Recruitment was done by the old regiments in the central provinces. The cadre was the regiment of Major General Bernhard Ludwig von Platen, taken over from the Duke of Holstein-Gottorp, with its 75 non-commissioned officers, 39 musicians and 635 privates. The Holstein officers went mainly to garrison regiments; they were replaced by Prussian officers from the army. This weak regiment, now a battalion, had taken Schwerin, seat of the Duke Carl Leopold, in the Imperial action in Mecklenburg, its 800 men being commissioned by the Emperor. Prussia represented Hannover and Braunschweig-Wolfenbüttel in opposing the Emperor. The Duke of Gottorp had become heir to the Swedish throne and had married the King's sister, Friederike Luise, in 1744. Since the regiment was intended for Silesia. its ranks were filled with Upper Silesian recruits. Its organization began in September of 1743 at Burg, its garrison until the King's death in 1786: Burg, Calbe, Genthin and Stassfurt until 1756, Burg alone from 1763 on, Crossen, Cottbus and Züllichau as of 1787, and as of 1789 in Glatz, with three companies in Silberberg. Replacements came from the start from the districts of Tost, Rosenberg and Lublinitz O/S and their cities, as well as Gleiwitz, Peiskretscham, Ujest, Guttentag and Landsberg. A straight-line distance of 510 kilometers had to be a disadvantage in terms of communications in those days.

In 1744 the regiment stayed with three others in the Berlin area, under the command of Prince Leopold of Anhalt-Dessau as of September 25. At the end of April 1745 he gathered a weak corps at Magdeburg, as Saxony was only pretending to keep its neutrality. After Hohenfriedeberg and their establishment of the situation in Upper Silesia, the King wrote to Prince Leopold: "Here we still have superiority over the enemy", and urged him to lead an attack on Saxony from the northwest. When the Saxon intention to advance on Berlin and Guben under Austrian protection became known on November 11, the King—after delaying for diplomatic reasons—ordered the advance. On December 15, on the left wing under Prince Moritz, it moved through the swampy valley of the Kessel Brook between Steinbach and Zöllmen to attack the heights, taking them in vigorous close combat. Peace was made ten days later. In 1747 the regimental commander, now a Major General, left Prussian service. The King's action against Austria was obvious, and Hesse-Darmstadt, unlike Kassel, was on the Habsburg side.

After being inspected by the King at the Pitzfuhl camp on June 17, 1756 it marched in the right wing of the army from Halle through Torgau, with the grenadiers guarding the food and bread wagons, to surround the Saxons at Pirna; then it occupied Dresden until October 16, 1757, at "middle foot" status with 1320 men, while the grenadiers took part in all the battles of that bloody year. Its officer corps embodied the spirit, conduct and moral strength of the army. Their renunciation, strictness of service, austere way of life, slow rise in rank, and sacrifice of welfare and life were regarded by the state only with esteem. Of all social classes, the nobility showed the best qualifications and determination. The order to the commandants of Dresden to enlarge it to 1800 men came to nothing. Only the grenadiers got their thirty extra men. In the Saxon Corps, the First Battalion took part in the advance to Hof. On June 25 the regiment was at Zschoppau, going then to Dresden, where it had been two years before. In 1759 it stayed with the Saxon Corps which advanced into Franconia at the beginning of May to weaken the Imperial army as much as possible before the campaign began. It turned aside to Bamberg and Nürnberg in order to reach Erfurt by the end of July. Because of moves that threatened their flanks, the corps had to turn prematurely and go to the Bautzen area. On July 30 I.R. 47 went to the Oder with the King to oppose the Russians, attacking Kunersdorf. Against the stubborn opposition of the Russian infantry on the Deep Path and under grapeshot fire from the big battery on the Spitzberg, it lost 600 men on August 12. Only two officers remained unscathed. Afterward it joined Finck's Corps, reaching Torgau in mid-September, winning the Battle of Korbitz on September 21, defending the area between Strehla and Torgau, and teaching the enemy a lesson in the fighting at Pretzsch on October 29. Finck was supposed to cut the enemy's communication lines to the rear, for which the King ordered him to march through Freiberg and Dippoldiswalde to the high plains at Maxen despite frost and snowfall. There he was quite on his own. On November 21 Finck surrendered after a heated battle in which he lost half his troops to more than twice their numbers; what remained of the I.R. 47 battalion was among the troops surrendered. In 1760 its one battalion was in Stutterheim's Corps, opposing the Swedes on the cker in Near Pomerania, then at Zehdenick north of Berlin, which it protected from October 5 to 9 under Duke Friedrich Eugene of Württemberg. At the end of October it passed through Magdeburg and joined the King, but stayed in Dben during the Battle of Torgau, where the grenadiers fought. In 1761, under Württemberg, it defended Kolberg, survived the breakthrough to the west on November 14, turned back to attack the battlements at Spie, and withdrew to Stettin, followed by Cossacks. At the end of December its attempt to surround the Swedes in Malchin failed. In 1762 its two battalions, serving with the army in Saxony, led the breakthrough over the Mulde on May 12, which developed into small battles. After returning on September 30, it did not take part in the Battle of Freiberg.

In Hohenlohe's Corps in 1806, it was disbanded by the Magdeburg capitulation of November 8. The Third and Fourth Battalions stayed in Glatz until peace was made in 1807 and joined with the garrison troops to form the Second Silesian Infantry Regiment, in 1914 the Grenadier Regiment King Friedrich III (Second Silesian) No. 11 of Breslau.

Füsilier-Regiment
Nr. 47

INFANTRY REGIMENT NO. 48
Fusilier Regiment

Commanders of the Regiment

1756	5/31	Lieutenant General Friedrich, hereditary Prince of Hesse-Kassel
1757	1/8	Colonel Friedrich Wilhelm von Salmuth, called Beringer, later Major General
1763	4/28	Major General Carl Friedrich von Beckwith, till then Chief of the "Prussian Legion" in the allied army of Duke Ferdinand of Braunschweig
1766	7/28	Colonel Matrin Ludwig von Eichmann, later Lieutenant General
1791	1/21	Lieutenant General Martin Ernst von Schlieffen, Governor of Wesel
1792	7/4	Major General Georg Bogislav von Köthen
1797	2/12	Field Marshal Wilhelm, Landgrave of Hesse-Kassel, since 1803 Elector of Hesse, Governor of Wesel

On June 28, 1755 the King ordered the Minden Garrison Battalion XIII Salmuth—founded in 1743 after the establishment of I.R. 32 as Wobeser, in 1746 Wutginau—to divide its companies, making ten from five, that is, to double it so as to be made into a field regiment later. It exchanged garrisons with I.R. 41 in Wesel, where it arrived on August 24 of that year with 13 officers, 25 non-commissioned officers, 13 drummers and 260 privates. Colonel von Salmuth learned on January 1, 1756 that I.R. 9, 10, 41, 44, 45 and the Garrison Battalion IX in Soest, Bielefeld, Minden, Wesel and Geldern were each to recruit sixty 'foreigners' and deliver them to Wesel by August 11, 1756. The rest were to be recruited by the paymaster captains of Geldern. The battalion was to exchange its tallest men or those from Minden for shorter men or those from the Wesel area, adding up to six officers, 25 junior officers and 350 privates. It also took over the uniforms of I.R. 41. The King gave command of the new regiment to the Hereditary Prince Friedrich of Hesse-Kassel, who simultaneously entered Prussian service as Lieutenant General and Lieutenant Governor of Wesel on May 31, 1756. Since he was not agreeable to taking over a garrison battalion, he was given I.R. 45 on January 8, 1757 and Colonel von Salmuth commanded the new I.R. 48, which had been at full strength, with 1400 men, since November of 1756. It had only one grenadier company, which was stationed in Magdeburg. Only in 1776 was it given a second, that of Garrison Battalion IX. Wesel remained its garrison; without a canton until 1788, it depended on recruitment in the region. In 1804 it was transferred to Paderborn, with three companies in Brakel and the grenadiers in Mühlhausen. Since 1801 replacements came from the Principality of Paderborn.

In 1756 I.R. 48 stayed in Wesel. When in December of that year the King sent Quartermaster General Count Schmettau to Hannover to set up an army on the lower Rhine and the Lippe, based in Wesel, it was kept there. For that reason, he planned from January of 1757 on to evacuate Wesel, which could not be held for a great duration. When England saw through France's desire for neutrality, it formed the Hanoverian Army, with a "Prussian Brigade", that April. It held the northwest flank open for the King, and included I.R. 48 on the "old foot" of 1220 men. After the evacuation of Wesel on March 24, it marched to Bielefeld, where it had a "sharp fight" with the French in mid-July, while its commander, Colonel von Salmuth, was defending Geldern. While marching across the Weser to Magdeburg at the King's command, it lost 320 deserters. Replacements from Magdeburg and Halberstadt brought it up to the "middle foot" of 1320 men. At this time the grenadiers were fighting at Prague and Leuthen.

On March 31, 1758 Prince Heinrich, to whose corps the regiment now belonged, had it filled, so it was ready to take part in the advance into Franconia in May, ending with its taking a position at Zschoppau on June 25. His corps was exposed to the pressure of the Imperial and Austrian armies, and on August 27 he defended the Gross-Sedlitz sector until the King reached Dresden in mid-September. Then it took part in Wedell's advance against the Swedes, who were threatening berlin and the taking of the Rhin sector on both sides of Fehrbellin on September 28, but the troops were needed in Saxony. When Daun's main army crossed the Elbe on November 7 to take Dresden, it strengthened the garrison under Count Schmettau, who refused to surrender and had the suburbs burned. On November 20 the King rode into Dresden and the enemies left. Back in the Saxon Corps in 1759, I.R. 48 again advanced to Bamberg and Ochsenfurt on the Main and could have gone on to the Danube. On May 23, under Schenkendorff, it attacked Austrian light troops at Aue and pushed them back. After the main Prussian forces withdrew to oppose the Russians, the Imperial army had an easy time in August: the First Battalion evacuated Leipzig, under Major General von Hauss, in return for free departure. The Second Battalion was in Dresden with five other battalions. Count Schmettau refused requests to surrender and prepared the defense. But a letter from the King, right after the defeat of Kunersdorf, gave him permission to evacuate the garrison, hospital, magazine and funds if necessary, as long as free departure was allowed, and led him to take that way out on September 4. The next day the King announced that he was coming. On September 26 the Second Battalion joined Finck's Corps, which led counterattacks, and went to meet the First Battalion, which had defended Torgau and, on the 21st, come through the Battle of Korbitz. Prince Heinrich turned back and joined Finck to hold the sector between Strehla and Torgau. At Pretzsch on October 29 the enemy took the next blow. In 1760 it reached the Dresden Heath with the King on July 10 and surrounded the city. But on July 22 it had to start marching back because Daun needed relief when the siege artillery failed to arrive on time. On August 20, under Hülsen, it withstood heavy attacks by three times its strength at Strehla, but fell back to Belzig and Beelitz by October 5. Saxony was given up. In 1761 it was with Prince Heinrich, who was able to hold his position on the Mulde cleverly. In 1762 it continued the small-scale fighting at the Elbe bridgehead between Dresden and the Erzgebirge. At Brand on October 15 it was thrown back to Freiberg by the Imperial army, with heavy losses, and almost captured. The victory at Freiberg restored conditions.

In 1806 it was disbanded in the capitulation of Erfurt on October 16, the Third Battalion by that of Hamelin.

Infanterie-Regiment Nr. 48

INFANTRY REGIMENT NO. 49
Fusilier Regiment

Commanders of the Regiment

1742	1/8	Major General Gerhard Cornelius von Walrave, sentenced to life imprisonment in 1748
1748	2/27	Colonel Philipp Loth von Sers
1758	11/26	Major General Christian Friedrich von Dierické, later Lieutenant General
1770	9/13	Colonel Carl August von Schwartz, later Lieutenant General and Commandant of Neisse
1788	6/1	Major General Otto Heinrich Friedrich von Borch
1794	12/29	Colonel Friedrich Wilhelm von Müffling, later Major General

After the taking of Neisse on October 31, 1741—focal point for the County of Glatz, northern Moravia and southern Upper Silesia—Major General von Walrave began to set up a ten-company regiment of engineers whose first job was the construction of fortifications. The men were recruited mainly in Upper Silesia, later also in Moravia, but the regiment was filled only in November of 1742. On January 8, 1742 most of its officers joined it from the Engineer Corps, and Walrave became its chief. Instead of grenadiers it was given two companies of miners. On November 25, 1741 the King gave Prince Leopold of Anhalt-Dessau the job of recruiting Harz and Halberstadt miners. Assembled in Kroppenstedt, they left for Silesia via Berlin on April 17, 1742. On March 30 of that year the King commanded that pontoon crews, counted as part of the artillery, be funded with the miners. Until then there had been engineers only in wartime. As of May 30, 1743 the construction of "Fort Prussia" in Neisse began; the "free watchmen" of the Silesian regiments were paid to serve there. The fortifications of Glogau, Brieg and Breslau were strengthened, as was Peitz. Glatz and Cosel were built up new. On the Old Dessauer's recommendation, Walrave was taken from the service of The Netherlands as a captain on November 24, 1715, and proved to be not only an expert on sieges and fort construction but also intelligent and imaginative, as he proved to be in Magdeburg, Stettin and, in 1733, at Philippsburg. A Colonel since July 2, 1729 and a Major General since 1741, he wrote a large treatise in 1747. What with his high regard for engineering art and military genius, the King promoted his knowledge until 1756 through the instruction of officers, including writing and practice. When personal dishonesty and "treason to the state" were discovered early in 1748, Walrave was sent to the "Sternschanze" in Magdeburg, where he was imprisoned until his death in 1773. The company of miners remained as an independent 'corps'. Only in 1787 was it given two companies of grenadiers from Garrison Regiment X. Neisse remained its garrison; from 1746 to 1750 part of it was stationed in Glatz. As of 1764 replacements came from the districts of Neustadt, Frankenstein and Grottkau (until 1797) and their cities, including Oberglogau, Zültz, Wartha and Steinau.

With the King's First Corps in 1744 it besieged Prague from September 12 to 16. Then it took Tabor in southern Bohemia and occupied it under Colonel von Kalnein. When the King turned back, it defended itself for days and cared for its wounded. On October 22 there was nothing to do but surrender. The regiment still had 661 men, plus 163 turncoats, most of them Austrian. After negotiations at Prague in December of 1744, the regiment returned in January of 1745. By now it was more fighting force than technical construction unit. Along with eight infantry regiments and the Silesian garrison regiments it secured forts in 1745, as well as in 1749 and 1753. In his 1752 report the King called it the "strong nail that holds the provinces together, the foundation of the army, nourishment of the troops, it protects the dwellings of great corps". For that reason he built forts in Silesia: Schweidnitz 1747-1752, Glatz as of 1742, Neisse since 1743 and Cosel from 1744 on. From 1746 to 1756 he spent 1.5 million Taler, excluding Silberberg from 1765. The sites in the old provinces were merely maintained as they were. He put special emphasis on establishing minefields and flooding perimeters. All forts were at once magazines and arsenals.

In 1755 the regiment exchanged their heavy swords for sabers; they stayed in Neisse through 1756. On January 9, 1757 it was on the 1320-man "middle foot" without being enlarged. After the defeat at Kolin the First Battalion was sent to defend the important magazine at Zittau, on the retreat route of Prince August Wilhelm's army. On July 23 it was captured in the burning city, with about 150 men under their commander, Colonel von Dierické.

In 1758 it joined the King's Army for the first time as a field regiment, used first in the siege and capture of Schweidnitz on April 16, then in the unsuccessful siege of Olmütz. On August 25 it stormed two enemy batteries at Zorndorf and capturted two howitzers and six powder wagons. Then it stayed in camp at Blumberg with the Pomeranian Corps under Dohna, opposite the Russians, while the King withdrew. On November 26 it was officially changed to a fusilier regiment when the miner companies were withdrawn, and Major General von Dierické became its chief. He came from an old military family, was well-educated and pious. On June 12, 1757 it was on the Warthe with the Pomeranian Corps, ready to take part in the planned advance to Thorn, which came to a stop at Obornik on July 5. Under Wedell at Kay on Kuly 23, it attacked the Palzig Heights for the fourth time in the second encounter and was beaten back, suffering severe losses. After the King and Finck arrived on August 9, it went to Kunersdorf on September 11, crossing the Oder at Gritz. On August 12 it protected the artillery there, holding out until the end when the enemy took the guns. The regiment was surrounded and most of it was captured after brave opposition. Those who escaped joined Manteuffel to push the Swedes back across the Peene at November and take Anklam and Demmin. At the end of April 1760 it held its position on the Triebisch under Prince Heinrich. At Breslau on August 6 and with the King in Saxony in mid-October, it attacked the Süptitz Heights at Torgau on November 3 and took several hundred prisoners. In 1761 it stood fast through hard times with Prince Heinrich in minor actions in Saxony. In 1762 it held the Pretzschendorf fort there as well as the area between Wilder Weisseritz and the Elbe.

In 1763 it received great favor, but in 1764 the King criticized it: "It doesn't amount to much either!" In 1806 it was disbanded in the capitulation of Magdeburg on November 11, the Third Battalion in that of Neisse on June 16, 1807. The remaining men joined the First Silesian Infantry Regiment, later Grenadier Regiment No. 10.

Füsilier-Regiment
Nr. 49
PRO GLORIA ET PATRIA
G. Dorn

INFANTRY REGIMENT NO. 50
Regiment on Foot

Commanders of the Regiment

1773	1/1	Major General Franz Ludwig von Rossieres
1778	11/23	Colonel Carl Ludwig Gottlieb von Troschke, later Major General
1794	12/29	Lieutenant General Georg von Steensen, Governor in Neisse
1799	10/1	Colonel Carl Wilhelm von Sanitz, later Major General

This musketeer battalion, the first to be established after the Seven Years' War, was intended by the King to occupy the Fortress of Silberberg, which was then under construction at the southeast edge of the Eulengebirge, ten kilometers west of Frankenstein. Under the direction of Lieutenant Colonel Regler of the Corps of Engineers, work had begun in 1765 on the construction of twelve connected forts blasted out of the rock on six hills at altitudes between 686 and 740 meters. As of 1779 the fortress cost the huge sum of 4.5 million Taler; it was never conquered. It was meant to protect the County of Glatz and the passes to Braunau and Nachod-Königgrätz, as well as the plains of Silesia. Colonel von Rossieres, who came from Sardinian service in 1765, was intended to be its commander. From 1766 on, Swiss and Württemberg officers were gradually taken into the service and trained with the Potsdam Guards until the battalion reached full strength at the end of 1768. The non-commissioned officers came from field and garrison regiments, already chosen individually as of June 1, 1772. The men came from the Silesian infantry regiments. As ordered on December 13, 1772, its actual establishment began as of January 1, 1773 in Reichenbach, where it remained until 1775, since neither Silberberg nor the battalion was ready. This peacetime status had lasted a long time, until the King's last enlargement of the army. It gained five musketeer companies with 810 men, 125 of them Prussian, with 685 'foreigners', and one grenadier company of 177 men, including 24 Prussians and 153 'foreigners', plus 25 officers, 59 non-commissioned officers, six medics and 21 musicians. After reaching full strength in April of 1775 it was sent to Silberberg in August, but only occupied the fortress in 1778, which was discontinued only in 1860. Here it stayed until 1796, with part of it in Glatz in 1795. In 1788 the one grenadier company was reorganized as musketeers, but in 1799 two new ones were formed in Münsterberg. With its transfer to Cosel in 1979 it gained its second and third battalions, becoming a complete regiment. The Third Battalion was stationed in Neisse until 1799. When the regiment served in Frankenstein from 1800 to 1806, the Third Battalion replaced it in Cosel. Though Cosel was also fortified, its special use until then as an immobile fort troop was thereby unnecessary, but in any case its move to Frankenstein made it a normal field regiment at last. Until 1796 its replacements came from the Silesian infantry regiments as reassignments, since it had no recruiting area. When it was brought up to full strength and moved to Cosel it gained the now-free canton of I.R. 42, which had been transferred to Warsaw: the districts of Beuthen, Pless and Gross-Strehlitz in Upper Silesia with the cities of Loslau, Nicolai, Myslowitz, Berun, Tarnowitz, Georgenberg and Leschnitz, but not the district towns.

In 1778 it took part in the Succession War in Bohemia, and from 1806 on it fought in the war with France as part of Hohenlohe's Corps. It was disbanded in the capitulations of Erfurt on October 16 and Magdeburg on November 11. The third battalion in the unconquered Fortress of Cosel joined the First Silesian Infantry Regiment in 1808, in 1914 the Grenadier Regiment King Friedrich Wilhelm II (First Silesian) No. 10 of Schweidnitz.

Infanterie-Regiment Nr. 50

INFANTRY REGIMENT NO. 51
Fusilier Regiment

Commanders of the Regiment

1773 1/1 Major General Otto Heinrich von Laxdehnen, formerly Commander of the First Guard Battalion, left Nov. 1.
1773 11/1 Major General Döring Wilhelm von Krockow, since 1786 Count and Lieutenant General
1789 2/7 Colonel Ernst Friedrich Carl von Hanstein, later Major General
1803 3/25 Major General Friedrich August von Kaufberg

After German border areas of Poland were separated and turned over to Prussia by the Treaty of Petersburg in 1772, after years of tension, the King established five new fusilier regiments in the newly gained Ermland and West Prussia as far as the Netze. They were his last. Funding for the first four regiments began on December 1, 1772, although they were not actually founded until 1774. Most of their officers came from the army, a few from foreign service, particularly subalterns. A group of officers from free battalions who had been pensioned off since 1763 became reactivated. There were 3,670 'foreigners' including non-commissioned officers; they went to the older regiments and were exchanged for people who had had less training and were intended to become non-commissioned officers. General von Lossow had 1370 of the necessary total recruited in Poland, also to trade for other 'foreigners' in Pomeranian, Magdeburgian and Westphalian regiments. The Westphalian Regiments 9, 10 and 41 were to provide 300 men, general recruiting was to bring in 1600, and Lieutenant General the Prince of Anhalt-Bernburg sent 400 from the Principalities of Anhalt, who were to be traded for other 'foreigners' via his regiment, I.R. 3. It was a somewhat laborious but carefully thought-out organization of personnel, quite comprehensible in view of the uneasy situation in the Polish neighboring areas.

I.R. 51 was founded in Marienburg on January 1, 1773, composed of recruits from the Marienburg and Danzig Werder areas, an area rich in German settlements, and put under the command of Major General von Laxdehnen. His replacements came from the districts of Marienburg and Dirschau, and later also from the suburbs of Danzig, plus the cities of Neuteich, Stuhm, Christburg, Putzig, Neustadt and Stolzenberg. In June of 1774 it moved into the Hochmeister Castle, a which had been rebuilt as barracks and was used as a vegetable storehouse after 1800. The inscription read: "Friedricus Magnus . . . ex ruinis restaurari iussit". It stayed in Marienburg until 1793 and was stationed in Danzig, which had become Prussian along with Thorn and Posen on January 23, 1793, from 1794 on. After this second partition Poland was no longer capable of existing, apart from the inner conditions. The Third Battalion, founded in Graudenz in 1796, followed it to Danzig in 1797, and was stationed in Neugarten, near Danzig, as of 1800. From 1799 on the grenadiers were also in Danzig. In 1778 it formed the avant-garde in Bohemia during the Succession War. In 1806 it fought in Württemberg's Corps, then under Blöcher, and surrendered at Ratekau on November 7; the remaining men joined the Third East Prussian Infantry Regiment, later Grenadier Regiment King Frederick the Great (Third East Prussian) No. 4 in Rastenburg. The Third Battalion took part in the successful defense of Danzig and joined the First West Prussian Infantry Regiment in 1808, later Grenadier Regiment Count Kleist of Nollendorf (First West Prussian) No. 6 in Posen.

Infanterie-Regiment
Nr. 51
Füsiliere

INFANTRY REGIMENT NO. 52
Fusilier Regiment

Commanders of the Regiment

1772	12/19	Major General Christian August von Lengefeld, later Lieutenant General and Governor of Magdeburg
1785	4/5	Major General Wilhelm Friedrich Carl, Count von Schwerin, later Lieutenant General
1795	5/19	Major General Joachim von Reinhart, later Lieutenant General
1807	1/30	Major General Carl Friedrich von Hamberger

The Cabinet Order of March 16, 1773 set the date for the founding of the new regiment for April 1 of that year. Its area of establishment was to be the Bishopric of Ermland, founded in 1243 and dependent on the Teutonic Knights until 1464, politically neutral since their decline but governed by the King of Poland, and now integrated into East Prussia because of its German population. The recruits were overwhelmingly Catholic. After the Hohenzollerns converted to Calvinism in 1613, on the advice of Count Abraham zu Dohna, religious tolerance prevailed in Prussia. The founding took place in Preussisch Holland (southeast of Elbing) and Mühlhausen (east of Elbing). Its commander was Major General Christian August von Lengefeld, who had been given command of I.R. 5 in 1785, came from the service of Württemberg, and had belonged to the King's suite for several years. The original cadre consisted of 'foreign recruits' who had already been assembled in April and May, as with I.R. 51. Junior officers were capable men from the garrison regiments. Enlarging the army always provided opportunities for advancement. The staff and subaltern officers came from other army regiments, including some 'foreigners'. Service in the King's army was attractive for many reasons and was regarded as an honor. By order of September 22, 1773, nine canton soldiers were placed in each company, to be replaced by nine more after two months' training, so that within a year a total of 54 men, in all practically a battalion, were trained as replacements of the most modern kind. As of April 15, 1774 all the canton men and those on furlough were called in for the usual two-month training period. Thus the regiment was fully ready for service within a year. Its garrisons were Preussisch Holland for the First Battalion and the two grenadier companies and Mühlhausen for the Second Battalion, which was transferred to Marienburg in 1796. In 1798 the whole regiment moved to Marienburg, in 1800 to Rastenburg and Rössel, three companies to Lyck, and the grenadiers to Angerburg in 1801, a move to the southeast because of the final partition of Poland. The Third Battalion with its musketeers, established in Mühlhausen in 1796 to take the place of the Second Battalion, was in Danzig in 1806. The recruiting area from 1773 to 1782 was the Ermland with the districts of Mehlsack, Braunsberg, Guttstadt, Wormditt, Tolkemit, Allenstein and Wartenberg, plus the cities of Elbing and Braunsberg. After 1783 replacements came from the districts of Marienburg, Heilsberg and Braunsberg and their cities, after 1795 from the newly East Prussian districts of Dombrow and Wigry and parts of the districts of Kalwary and Bialystok, with nineteen cities, in the area south of the provincial boundary.

Serving in Bohemia in 1778 during the Succession War, it marched as the avant-garde and took part in the Battle of Zuckmantel on January 14, 1779, and was accorded the Grenadier March. In June of 1780 First Lieutenant von Moeck was given the Pour-le-merite because at Wartha on February 14, 1779 he and his forty men fought off an attack of 400 men on his 200-wagon food transport train underway from Frankenstein to Glatz and, without losing a man, had brought in ten prisoners. In the Battle of Rawka on June 6, 1794 against the Poles under Kosciuszko it did so well that on June 8 of that year Colonel von Nastzmer, the Regimental Commander, and the three Battalion Commanders, Lieutenant Colonel von Obernitz and Majors Johann Friedrich and Matthäus Bernhard von Krasewski, received the pour-le-merite. Next it besieged Warsaw. In 1806 and 1807 it took part in the defense of Danzig under General von Kalckreuth until May 24 and was allowed free departure; the Third Battalion under General von Courbiere was still in service on December 12, twenty-two weeks after the armistice.

In 1808 it formed the First West Prussian Infantry Regiment, in 1914 Grenadier Regiment Count Kleist of Nollendorf (First West Prussian) No. 6.

Füsilier-Regiment
Nr. 52

INFANTRY REGIMENT NO. 53
Fusilier Regiment

Commanders of the Regiment

1773	5/22	Colonel Caspar Fabian Gottlieb von Luck, later Major General
1780	6/9	Colonel Wilhelm Heinrich, Baron von der Goltz, later Major General and Commander of I.R. 7
1784	4/5	Colonel Wilhelm Friedrich Carl, Count von Schwerin, transferred to I.R. 52
1785	4/5	Colonel Carl Albrecht Friedrich von Raumer
1786	3/1	Major General Franz Andreas von Favrat, transferred to I.R. 33, later Governor of Glatz
1794	12/29	Major General Albert, Count of Anhalt
1800	10/2	Colonel Wilhelm Christoph von Larisch, later Major General

On May 22, 1773 the regiment was given Colonel von Luck as its chief; on June 1 it was stationed in Braunsberg, midpoint of the Ermland, a triangular area with each side about a hundred kilometers long, reaching from the source of the Alle to the western edge of the Masurian Lakes. The period of almost three hundred years in which this distant province had been separated were finally over. Thousands of men from the 356,000 Germans recently regained now came into close contact with the Prussian Army for the first time. They became loyal, dependable soldiers, as long as order and righteousness were valued and any Prussian nationalization was not mentioned. The core of officers, junior officers and privates came from old regiments. By the end of 1773 the 'foreigners' recruited from the Empire were all gathered. The introduction of recruits from the Ermland began in January of 1774; their two-month training in successive groups took place as with I.R. 51 and 52. Thus the time needed to complete the regiment was relatively short. Like the two other regiments, it went to its two-month summer training as early as April 15, complete with canton soldiers and those back from furlough, and was then complete and ready for action. The first three new regiments between Marienburg and Braunsberg were completed at the same time; the two West Prussian ones on the Vistula came only later. Until then there had been no more than five infantry regiments in all of East Prussia. Conditions in the outside world were stormy enough. The First Battalion was garrisoned in Braunsberg, the Second in Heiligenbeil until 1777. After serving in Bohemia during the Succession War in 1778, the whole regiment was stationed in Braunsberg. After the 1794-1795 Polish campaign, the regiment was sent to Thorn in West Prussia. The grenadiers were in Preussisch Stargard. The musketeers of the Third Battalion, founded in Gollup in 1796, were in Königsberg from 1797 to 1799, in Hohensalza from 1800 to 1805, in Graudenz in 1806. Replacements came from the districts of Tapiau and Brandenburg/Pr. with the cities of Kreuzburg, Heiligenbeil, Wehlau and Tapiau until 1798, when the canton became the district of Lipno, part of the district of Mlawa, Wyssogrod, Heilsberg, Seeburg, Rössel, Kowalewo, Gollup, Strasburg, Löbau and Michelau, and the cities of Gollup, Strasberg, Leutenburg, Kauernick, Rypin, Kikol and Plock. The list of commanders includes numerous qualified officers.

In 1778 it marched under Lieutenant General von Wunsch to Nachod in Bohemia. It was mobilized in September of 1794 to put in effect the treaty signed in St. Petersburg on January 23, 1793, and fought at the bridge of Segrz on August 18 and on the Narew on August 24, standing firm against attacks of Dembe, Jachronka and Sierok, as seven Pour-le-merite prove. The names of prominent Polish families turn up among the captains. The regiment was accorded the Grenadier March. In Württemberg's Corps and then Blöcher's in 1806, it was disbanded in the capitulation of Ratekau on November 7. The Third Battalion in Graudenz joined the First West Prussian Infantry Regiment, later Grenadier Regiment Count Kleist of Nollendorf No. 6; the garrison soldiers went to the Prussian Artillery Brigade.

Infanterie-Regiment
Nr. 53
Füsiliere

INFANTRY REGIMENT NO. 54
Fusilier Regiment

Commanders of the Regiment

1773 12/1 Colonel Albrecht Ehrentreich von Rohr
1784 4/22 Colonel Carl Kuno Ludwig von Klitzing, later Major General
1786 3/1 Major General Bogislav Ernst von Bonin, later Lieutenant General
1794 12/29 Major General Christian Friedrich von Mosch, later Commandant in Wesel
1799 1/28 Colonel Hans Christoph von Natzmer, later Major General

The fourth new fusilier regiment was founded in Graudenz and Culm as of December 1773 and put under the command of Colonel von Rohr on December 1. The officers came in part from the army, in part from foreign service. Junior officers were transferred from the Magdeburg Inspection, and privates were sent from the regiments on the border of East Prussia and Silesia. The number of 'foreigners' recruited was complete only in September of 1774. The summoning of one-third of the canton troops at a time—the method of short training seemed to have worked out—for a period of two months began on command as of March 17, 1774, after only fourteen weeks, "so that the training of the entire mass will not be delayed". The canton soldiers, now mainly Polish, were to be replaced through recruiting in the Empire. The least dependable recruits were not furloughed either, but stayed with the troop to strengthen the effects of training and comradeship. The regiment was located in the "Kulmer Land" between the Drewenz and the Vistula, in the area of the cities of Kulm, Bromberg and Thorn, whose name was derived from "Kölmer" (in Polish, "cholm": hill). Since 1773 this area was the new "West Prussia", while the Province of Prussia was now called East Prussia. In 1772 the King had established a new war and domain council for the area in Marienwerder. It formed the important land-bridge to Pomerania and Brandenburg, 125 to 275 kilometers wide. For this area the immediate availability of, in all, about eight thousand men was certainly not excessive. From 1773 to 1777 the garrisons were Graudenz for one battalion and Culm for the other and the grenadiers. From 1779 to 1792 the whole regiment was garrisoned together in Graudenz. From 1795 to 1806 it was again divided between Graudenz and Culm, as before; the grenadiers were in Marienburg since 1799. The Third Musketeer Battalion, founded in 1796, was in Graudenz and Marienwerder until 1799, and from 1800 on likewise in Graudenz, which developed into a strong garrison. Replacements came from both sides of the Vistula, from the districts of Culm, Culmsee, Lippinken, Roggenhausen, Rehden, Graudenz, Schwetz and Tuchel with the cities of Culm, Graudenz, Rehden, Briesen, Schwetz, Konitz, Lessen, Kowalewo and Culmsee.

Along with the foundation of new regiments, the King had continued from 1774 to 1777 to strengthen the old regiments, which he had started and finished for the time being in 1768 and 1769. In the Succession War in 1778 the regiment belonged to the King's First Army and was stationed in Bohemia from July to September, then in Upper Silesia. On April 24, 1794 it marched along into Poland after the Russians had arrived in 1792, made the Petersburg Treaty on January 23, 1793 and faced a Polish uprising. At the end of May it was at Lowicz and Rawa; the grenadiers fought at Skala, not far from Krakau, and on the left wing in the Battle of Rawka, near Chebdzie, on June 6. From June 13 on, the regiment was stationed at Wola, west of Warsaw, to begin the siege. For the storming of the battlements of Wawrzyskew and the Bastion of Powonske on August 26 and 28 it received four Pour-le-merite on September 3. The soldiers of the Culmer Land had handled themselves well in this psychologically difficult situation. In Württemberg's Corps and then Blöcher's in 1806, it was disbanded in the capitulation of Ratekau on November 7. Men on garrison duty went into the Fourth East Prussian Infantry Regiment; the Third Battalion, which had helped to defend Graudenz, joined the Second West Prussian Infantry Regiment in 1808, later the Grenadier Regiment King Wilhelm I (Second West Prussian) No. 7 of Liegnitz, the so-called "King's Grenadiers".

Infanterie-Regiment
Nr. 54
Füsiliere

INFANTRY REGIMENT NO. 55
Fusilier Regiment

Commanders of the Regiment

1774	2/23	Colonel Adolph, Prince of Hesse-Philippsthal, later Major General
1780	6/18	Colonel Georg Ewald von Blumenthal
1784	9/24	Major General Melchior Sylvius von Koschenbahr
1790	9/26	Major General Carl Eduard von Thiedemann, ex-I.R. 7
1792	1/26	Major General Heinrich, Count von Brühl
1792	4/1	Colonel Carl Friedrich von Holwede
1796	11/4	Major General Ernst Johann von Manstein, later Lieutenant General

The fifth fusilier regiment appeared on the budget only in June of 1773. On February 23, 1774 the King turned it over to the Prince of Hesse-Philippsthal. Its organization proceeded fairly slowly. It was formed in Mewe on the Vistula and Preussisch Stargard to the west, thus in Pommerellia, which belonged to the Teutonic Knights from 1308 to 1569 and was called the "Polish Corridor" from 1919 on. A small number of its officers came from the army, most from foreign service. The non-commissioned officers came from the Westphalian Inspection. The officers met in May of 1774. As of the end of June the canton soldiers from the area could be called in at the rate of thirty men per company for two-month training and then release. At the same time it was allowed to recruit additional men in the Empire, since the majority of the canton soldiers were Polish. As in I.R. 54, the least dependable of them were not furloughed, but kept with the regiment. The cadre of 'foreigners' was complete only in January of 1775. The short two-month training period was used in the cavalry regiments in the same way. After the regiment was prepared, it moved to the West Prussian Infantry Inspection under Colonel von Rohr, Commander of I.R. 54. I.R. 4, 51, 52, 54 and 55 belonged to it. One battalion each was garrisoned in Preussisch Stargard and Mewe from 1774 to 1795, except for 1775, when the whole regiment was in Mewe. In 1796 it was divided among Bromberg, Gnesen, Hohensalza and Strelno, thus moving farther south into the Province of Posen, which had become part of Prussia in 1795. In 1797 Strelno was discontinued, as was Hohensalza in 1800; the grenadiers were in Preussisch Stargard as of 1799. Recruits came from the districts of Konitz and Preussisch Stargard and parts of the district of Dirschau, as well as the cities of Konitz, Friedland, Hammerstein, Baldenburg, Tuchel, Schlochau, Landeck, Berent and Dirschau. The growth of West Prussia had finally put the defense of East Prussia on a firm foundation and made the Vistula a possible line of defense. In 1776 the King ordered the construction of a fortress at Graudenz, which was to play a role later.

In May and June of 1778, before the Succession War began, it was quartered between Münsterberg and Neisse, in order to be available to the King's Army for the campaign in Bohemia from July 5 to September 21. From September 30 on it was in the Hereditary Prince of Braunschweig's Corps in Upper Silesia. When it was mobilized on November 8, 1792 it called its furloughed men in but stayed in garrison, being intended to take part in the occupation of Thorn, which began only on May 7, 1793. On April 24, 1794 it was mobilized and sent on the march. By the end of May the First Battalion was in the Lowicz-Rawa-Petrikau area, the Second and the grenadiers at Tschenstochau. Since 1762 the Russians and Prussians had been fighting on the same side for the first time. In the Battle of Rawka on June 6, 1794 the Second Battalion and the grenadiers fought on the left flank near Chebdzie, called "Seelze" in Prussian documents. On July 13 the whole regiment was in the King's Army before Warsaw.

On August 26 and 28, under the command of Major General von Pollitz, it charged four battlements at Wawrzyszew and the Powonsk outworks with bayonets, capturing six cannon and two howitzers. The entire Polish right wing was thrown backward. The King ordered a general attack on Warsaw for September 1, which was made impossible by defeats and differences of opinion with the Russians. Between August 29 and November 25 the regiment received ten Pour-le-merite for its achievements. At Sochaczew the grenadier battalion fought for hours on October 19 and 23 against ten-to-twenty fold superior forces "with the greatest bravery". Its commander received a letter from the King and early promotion. In Württemberg's and then Blöcher's Corps in 1806, it surrendered at Ratekau on November 7. Garrison troops joined the Third East Prussian Infantry Regiment, while in 1808 the Third Battalion of Graudenz went to the Second West Prussian Infantry Regiment.

Füsilier-Regiment
Nr. 55

DRUMMERS OF REGIMENTS NO. 1 AND 5

To maintain marching order and the even, uniform progress of battle movements, often across the fields, the drummers or "Tambours" were rhythmically indispensable. Then too, troops were led orally and optically in those days, particularly since the Old Dessauer had introduced marching in step. In the clash of firmly controlled troops in a small space, shooting with strong-smelling black powder and fighting man-to-man on a broad front, spoken commands were no longer possible. But in other ways too, the daily course of the soldier's life was accompanied by the sound of drums: from waking up to going to bed, in drill and guard duty, presenting, assembling, going to church, flag march and alarm (general march), in recruiting, for singing and dancing, even at burial. The drums were silent during loading, but their subdued storming march inspired unity, courage, a forward-moving intensification to the fury of the attack, thus a psychological effect. The "Grenadier March", a march to drums only, for foot soldiers, was granted by the King as his highest recognition of bravery in battle, but could also be withdrawn if the troop failed to do its job.

Until the early Eighteenth Century, military music had been limited for centuries to drums and pipes and remained almost unchanged. Toeche-Mittler calls the Eighteenth Century "the century of the musicians". They gained significance with the rise of standing armies. The name "Tambour", a name derived from Arabic, was at first an instrument, a type of percussion guitar called "hand drum" in French (compare with "Tambourine"), but since the Seventeenth Century meant the drummer himself. When French terminology was taken over for military things, the use of the Middle High German term "Drummelschlger" (drum-beater) became limited, while "trummel" (drum) was retained from 1655 on. Its name means nothing more than "noise-maker" with rhythm and accompaniment, just as "Schlagzeug" (percussion) implies today, with an amplifier. The regional soldiers of old already had two "players" per flag, a drummer and a piper. The drums were made of sheet metal, manufactured in Berlin since 1698. There were no written notes, for the drums were used daily and passed on, with training in their traditional use, from man to man.

All drummers, fifers, trumpeters and oboists in the Prussian Army caught the eye because of their coats decorated with braid or other trimming, especially on the sleeves down to the wrists. They were recognizable from afar, and their arm movements were noticeable as well. Bleckwenn indicates that the musician's coat with its "swallows' nests", stripes across the sleeves, seam and corner trim, were originally "servants' livery" when in the Seventeenth Century the musicians were personal servants of the company commander. For that reason the decorations were different in every regiment and incomparable, for their bright colors, often patterned on heraldic designs, were the colors of the regimental commander's coat of arms, those of the man who, at the beginning of the Eighteenth Century, had personified or even founded the regiment, and so they had remained. Only the reorganization of the old Prussian army in 1808, in times of greatest privation and a new sense of style, did away with these splendid uniforms. Only the "swallows' nests", which had once developed out of shoulder padding, retained their characteristic braid trim until 1945 as a symbol of the military musician. Only one original coat has survived to the Twentieth Century.

Every musketeer company had three drummers, a battalion as a whole had sixteen. The oldest companies also had a fifer. The regimental "Tambour", outranking 32 drummers, was the senior drummer of the first battalion. The grenadier companies likewise had three drummers, but also two fifers per company. When the two grenadier companies were with the regiment in peacetime, there were 37 in all; when in wartime a grenadier battalion was formed from two regiments, it had twelve drummers and eight fifers. In the lower staff of the regiment there were six fifers and six oboists. It must have been a unique type of music, with so few bassoons and oboes as melody instruments up against such powerful percussion! The trumpet was the signaling instrument, the trumpeter the "choir director". The drum rings were also decorated in unique ways, the colors corresponding to those of the regiment or the decorations of the musicians' uniforms. The Guards had red-and-white zigzag drum rings, which King Friedrich Wilhelm III made uniform for the whole army on January 28, 1817.

The oboist took his name from the beechwood or ebony woodwind instrument that was called the "French shawm" in the Seventeenth Century. When standing troop musical units were instituted after 1650, the then most popular wind instruments, the oboe and bassoon, were chosen to play the melody and thus became the main instruments. The designation spread to all infantry musicians, though from 1770 on clarinets were actually the main wind instrumenmts. The oboist wore the hat of a junior officer and the white sash of the bodyguard company, and belonged to the junior staff of the regiment. The regular drummer's braid on his coat was often accented with metal thread. But the musician received only the a private's pay. Their esteem in the troop varied; they were not regarded as steadfast, although, for example, in 1808 the oboist Stellien rescued the flag of the First Guard Battalion from the enemy at Prenzlau.

1. Rgt. 5. Rgt.

The "drummer boy" of olden days, who was also a servant of his captain, disappeared only in Frederick's day. Drummers could not have been too small and weak, for a sheet-metal drum weighed considerably more than the wooden drums of earlier times. The captains were interested in taking boys into their personal service at first as new recruits, new growth, and then using them as drummers, having them enter the rank and file in cases of vacancy. They had taken their oath and gone along to battle, but were not yet trained in the use of weapons. Naturally they were often teased crudely on the march or in camp, sometimes made fun of out of a long-term soldier's desire for prestige, or simply out of boredom. The later responsibility of the military musician to work as a stretcher-bearer or assistant medic did not yet exist then. As Bleckwenn reports, the regulations of 1726 banned it, so as not to weaken the troop in battle or give any excuse for disappearing to the rear. Since the battles lasted a considerably shorter time, care for the wounded was possible afterward. Modern concepts of the situation cannot be applied to those times.

The "Librery" (French "livrée", also in the sense of "external identifying mark") of the musicians was determined by right of the commander, as the emblem of the colonel and chief of the regiment at the time, as well as the flags, bullet cases, hatbands, grenadier caps, and in the cavalry, saddle cloths, kettledrum straps, trumpet cloths and braid. The "drummer's cords" in the colors of the regimental commander's coat of arms were applied to the front, back and sleeve seams and the pockets for decoration and reinforcement. They also filled in the sleeve surfaces with transverse lines or chevrons and seamed the "wings" or flying sleeves hanging from the shoulders, which were customary in The Netherlands since the beginning of the Seventeenth Century. Around 1700 these "wings" went out of style in the infantry. The colors of the drummers' coats usually differed from those of the soldiers; often they were the opposite colors, for example, in the Anhalt Regiment in 1683, red "with wings and blue and white cords", and in the Electress Regiment, which wore red coats, blue with red and white cords. The wooden drums still bore the coats of arms; the sheet-metal drums had only the colors on the rims. Drumsticks were worn over the right shoulder on the march, inserted in the eyes of thongs fixed to the drum-strap. The fifes gave way to shawms, until they in turn were replaced by the melodic French oboes. The musicians originally had "choruses", a term later limited by the use of the term "corps". According to Dr. Hans Bleckwenn's research, the trimmings of drummers' and other musicians' coats had the following significance:

Infantry Regiment 1:
Figures and colors are taken from the coat of arms of Major General Caspar Otto von Glasenapp, General of the Infantry in 1740, later Field Marshal, Regimental Commander from 1723 to 1742.

Infantry Regiment 2:
The three colors and the tassel of the hat come from the arms of Major General Erhard Ernst von Roeder (1665-1743), not only a good drillmaster, but also highly esteemed by the King "because of his education, his clear sight and his great energy one of the most capable higher generals", 1728 Inspector in East Prussia, 1733 Commanding General in the War of the Polish Succession, 1739 Field Marshal.

Infantry Regiment 3:
The simple white trim is typical of the regiments of the House of Anhalt, to which it belonged between 1670 and 1784 except for a period of fourteen months, under five princely commanders.

Infantry Regiment 4:
The colors belong to the arms of Colonel Georg Rudolf von Glaubitz, Regimental Commander from 1731 to 1740, who then, as a Lieutenant General, commanded Garrison Battalion 3.

Infantry Regiment 5:
Until 1754 they were the colors of Lieutenant General Christoph Heinrich von der Goltz, 1720-1732 Commander of I.R. 15, then chief of this regiment until 1739, afterward the house colors of the House of Braunschweig.

Infantry Regiment 6:
Here the colors of the King were used.

Infantry Regiment 7:
Colonel August Wilhelm, Prince of Braunschweig-Bevern, made some changes when he took over the regiment in 1741, but refrained from using his house colors in the trim, perhaps out of consideration for the Brandenburg tradition.

Infantry Regiment 8:
Here the colors are those of the arms of Anhalt from the 1690-1747 period.

Infantry Regiment 9:
The ring in the trim is said to come from the Lattorp arms, but only in form, not in color.

Infantry Regiment 10:
Red and white are the colors of the House of Hesse, 1703-1730.

Infantry Regiment 11:
This trim has no special personal or heraldic tradition.

Infantry Regiment 12:
The trim bore the Hessian house colors since 1752, in addition the blue of the lion shield.

Infantry Regiment 13:
The trim shows the colors of the Regimental Commander from 1715 to 1722, Lieutenant General Anton von Pannewitz.

It is noteworthy that most of the colors originate in the period of peace between 1715 and 1740, when the foundation of the army was formed and established.

Inf.-Rgt. Nr. 11

Inf.-Rgt. Nr. 13

Commanders are the fate of their troops; that has been true without limit for centuries and is often true of commanders in Frederick the Great's armies. What they left behind in terms of training, spirit, and unity of the staff lived on for years or even decades, naturally often determined by chance, in which military history is rich. The aftereffects on flags, symbols and emblems for uniforms were very typical of this era and testify to the existential link between chief and troop.

Here the colors continue that Dr. Bleckwenn has determined:

Infantry Regiment 14:
Red is the King's color and orange the symbol of the Prince of Orange, the title given by King Friedrich I of Prussia in 1708 and 1712 to the regiments commanded by his son, Crown Prince Friedrich Wilhelm I, the title closely linked to the royal family of The Netherlands.

Infantry Regiment 15:
As of 1740 the new King's Guards naturally show red in all three battalions, but also silver as a sign of high position and worth. The Unranked Corps wore only red and white.

Infantry Regiment 16:
The colors black-white-red come from the flag of Field Marshal Adam Christoph von Flanss, Regimental Commander from 1728 to 1748, as of November 16, 1743 Supreme Commander in East Prussia until his death in 1748. He became a Field Marshal on January 17, 1745.

Infantry Regiment 17:
The silver-red-green colors come from Friedrich Wilhelm von Grumbkow, Field Marshal and Minister, colleague and friend of King Friedrich Wilhelm I. After his death in 1739 they were replaced by green and yellow in memory of his predecessor, Colonel Baltzer Friedrich von Sydow, Regimental Commander from 1693 to 1703.

Infantry Regiment 18:
Red and white, very common as colors of arms, apply equally to Lieutenant General Anton von Pannewitz, Major General Daniel von Tettau, and Lieutenant General David Gottlieb von Gersdorff between 1698 and 1731. Completion with green comes from Major General von Derschau, Regimental Commander from 1738 to 1742.

Infantry Regiment 19:
The eight-pointed Cross of St. John in the color of Christ, white, is the emblem of the order and stands on a red field, the color of the Holy Ghost. After Lieutenant General the Prince of Braunschweig the colors became white with yellow stripes.

Infantry Regiment 20:
The quarterings and color combinations white-blue-red come from the arms of Major General Johann Heinrich von Börstel, who was closely linked to the regiment and its core of troops from 1698 to 1711. The tassel recalls his successor, Major General Ulrich Christoph von Stillen, Commander from 1711 to 1728.

Infantry Regiment 21:
The white stripe was originally framed in blue, now with yellow borders on both sides. The colors come from the arms of General of the Infantry Heinrich Karl von der Marwitz, Regimental Commander from 1724 to 1744.

Infantry Regiment 22:
Plain white is the tradition of the House of Anhalt-Dessau. The youngest son of the "Old Dessauer", Field Marshal Moritz, Prince of Anhalt-Dessau, was Regimental Commander from 1741 to 1760.

Infantry Regiment 23:
There is no explanation for the white trim of this regiment with the striking blue web stars and the lack of sleeve decoration—only swallows' nests—although the regiment was a famous one.

Infantry Regiment 24:
The red diamonds in the white and yellow bars come from the arms of Field Marshal Curt Christoph von Schwerin, Regimental Commander from 1723 to 1757.

Infantry Regiment 25:
The dark red stripe with the black diamonds refers to the arms of Lieutenant General Anton von Pannewitz, who as a Major General was Commander of the Second Battalion from 1713 to 1715.

Regiment Nr. 15, III. Btl.

Regiment Nr. 18

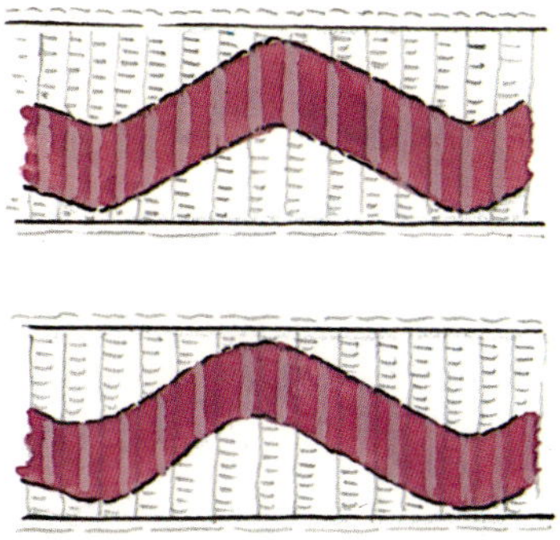

DRUMMERS OF FUSILIER REGIMENTS NO. 43 AND 46

Except for I.R. 44, 45 and 48, the drummers and oboists of the regiments from I.R. 31 upward wore no sleeve decorations any more, but only the musicians' decorations of swallows' nests, edging on the cuffs, on pockets, trouser seams and front coat seams. From I.R. 28 up they had formerly been fusilier regiments of the Soldier King, which Frederick the Great turned into "Regiments on Foot", thus to musketeers. From I.R. 32 on they were new fusilier regiments established by him.

Introduced in France in 1640, the "fusil" was the new grapeshot gun, as opposed to the traditional flintlock gun, allowing faster fire, later a shorter, lighter gun that was simpler to load as a muzzle-loader and did not shoot so far. One could also attach a bayonet to the barrel, later next to it, so that one could shoot and stab at the same time. This made the pike superfluous as a weapon. There were fusiliers in Prussia since 1723. They differ from musketeers only by their fusilier caps, made to imitate the grenadier cap with a smaller visor and point on the crossed bows. Bleckwenn calls them a "naive attempt to deceive". In Prussia they were at first regarded as second-class, while in all the rest of Europe they remained an "elite troop"; later they were seen as an exemplary troop, and after 1787 as light infantry for loose, scattered fighting; in the Twentieth Century their name is merely a tradition.

The drummer of I.R. 43 wears white with an orange chain of diamonds. Since the regiment originally consisted of former Breslau city soldiers, it is possible that the Breslau city colors of yellow and red, which were formerly its symbol, are continued here. The drummer of I.R. 46 shows the yellow-red colors of Württemberg in the yellow stripe with red diagonal bars and black and red figures, symbolic of Duke Carl Eugene of Württemberg, its chief from 1743 to 1757. They make a nice, effective contrast to the black facing and appliques.

According to Bleckwenn, the colors of the musicians have the following significance:

Infantry Regiment 26:
Two red foxes running on white, separated by a red diagonal; both color and figures refer to the arms of von Kleist (1730-1749).
Infantry Regiment 27:
The plain white trim—as with I.R. 3 and 22—denotes the House of Anhalt.
Infantry Regiment 28:
The first early fusilier regiment, without sleeve decorations; the deep red decoration has no heraldic significance.
Infantry Regiment 29:
No suitable explanation has been found for the broad white band with double red-blue stripes from outside to inside.
Infantry Regiment 30:
The white trim with the yellow zigzag lines in the middle of a double blue stripe probably comes from the arms of von Sydow (1717-1731; first garrison, then fusilier battalion).
Infantry Regiment 31:
The pink stripe with white double corner, small yellow diagonal stripes and white diamonds are reminiscent of the arms of von Dossow with their three roses (Commander 1729 to 1743).
Infantry Regiment 32:
The diamonds in red, surrounded by white, in a blue stripe with a yellow border are meant to combine the arms of Roeseler (1725-1738) and von Schwerin (1743-1747).
Infantry Regiment 33:
There is no explanation for the blue band with white borders and white double stripes; see the flag.
Infantry Regiment 34:
The yellow-red band with additional blue middle stripe refers to the Crown Prince Regiment, where the middle stripe has been added.
Infantry Regiment 35:
There is no explanation for the blue-yellow checkered band.
Infantry Regiment 36:
The white stripe with blue corners has no relation to the arms of von Münchow (1740-1766); compare the white decorations and colors of the grenadiers.
Infantry Regiment 37:
No explanation is known for the white decoration with red-rimmed blue double stripes.
Infantry Regiment 38:
Likewise no reason for the white band with three red stripes is known.
Infantry Regiment 39:
The silver stripe with the yellow pierced diamonds, halved at the edges, refers to the colors of the House of Braunschweig, to which the regiment belonged from 1740 to 1771.
Infantry Regiment 40:
The pink band with white wavy line and corners is as yet of unknown origin but strikingly resembles the flag.
Infantry Regiment 41:
The plain yellow trim—identical to No. 50 in the edging—has not been explained.
Infantry Regiment 42:
The white band with the orange wavy line, the same color as the flag, indicate the Margraves of Brandenburg-Schwedt, related to the House of Orange, which used these colors (1741-1789).

Infantry Regiment 44:
The full yellow edging with red corners, with black stars inside and white center, surely refers in terms of its pattern to the moor's head in the arms of Jungkenn Müntzer von Mohrenstamm (Commander 1749 to 1759).
Infantry Regiment 45:
The sleeve edging in white with red corners and diagonal chains of flowers alternating with blue diagonal bars corresponds to the arms of von Dossow (granted 1743, Commander 1743 to 1757).
Infantry Regiment 47:
The white stripe with red corners and the same pattern as I.R. 46, only in blue, refers to the colors of the House of Hesse-Darmstadt.
Infantry Regiment 48:
The full sleeve edging with brown and white corners and brown diagonal bars comes from the von Bredow arms.
Infantry Regiment 49:
The orange band with blue corners and wavy line has no known source.

Füs.-Regt. Nr. 43

Füs.-Regt. Nr. 46

GRENADIER CARPENTER OF INFANTRY REGIMENT NO. 3

If the grenadiers were storming battalions, then their carpenters were "maids of all work". Every grenadier company had six, later even seven carpenters, making twelve per infantry regiment, but 24 per grenadier battalion. As a rule their place was with the regimental flags. The grenadiers of the garrison regiments had the same number. As regimental engineers they built roads, improved bridges, removed obstacles, fortified positions and overcame problems. In 1748 the King decided to call on the grenadier-carpenters to man the 236 battalion cannon—every battalion had two three-or six-pounders—as had already been done at Hohenfriedeberg. Since then the carpenters were always with the artillery every summer for training in loading and shooting, for example, as of May 7, 1749 in Glogau, Breslau, Brieg, Neisse and Glatz. The artillery could use practicing handworkers. Both developed from craftsmens' guilds and had learned to use explosives. So the carpenters became a sort of "second artillerymen". But on the march—often enough across the fields—they built crossings or causeways ("swamp bridges") and showed their engineering capabilities again and again.

On February 1, 1750 the King had the number of carpenters with the grenadier companies increased to seven, for a total of 819. In war they always joined the first company of a musketeer battalion. In exchange, most grenadier battalions received sixteen artillerymen, and the two musketeer battalions an additional nine artillerymen, so that every battalion cannon had an eight-man crew. Now the grenadiers had better-trained gun crews. As a rule, in every infantry battalion a junior artillery officer directed the battalion's cannon. The artillery-carpenters lost their particular uniforms from 1754 on and were integrated as artillerymen.

Typical of them were the brown leather apron and the axe, which was standard for the whole army. It was carried in the bullet-pouch, in which no bullets were kept. They carried their gun bullets in the small pouch at the front of their belt. As commanded on August 25, 1753, the guns, which were a hindrance, were to be given up. But various reports leave the question open as to whether this was carried out everywhere. In any case, the carpenters were important and interesting soldiers.

OBOIST AND FIFER INFANTRY REGIMENT NO. 31

Among the lower staff of the regiment were six fifers which only the infantry had. The fife already existed in the days of the Sixteenth-Century mercenaries, when it was called a "drummer's pipe". Its high range and sharp accentuation formed an interesting contrast to the originally dull, later ever-clearer sound of the drums. The fife was introduced into Prussia under Friedrich Wilhelm I in 1714. In the Frederician army the fife was not uniform in all troops; there were various lengths and patterns. Only the grenadiers actually had fifes, with twenty drummers and fifers per battalion. There were also two so-called "musketeer fifers" in the bodyguard company and the company of the commander of the second battalion. On July 30, 1749 the King made their places free for carpenters in his plans. In 1755 he planned to do the same for grenadier companies of the garrison troops, but nothing seems to have come of this. In uniform and rank the fifers were equal to the drummers.

Zimmermann

Inf.-Rgt. Nr. 31

Hoboist

Pfeifer

DIVISION BY STATES OF THE INFANTRY REGIMENTS, 1740 TO 1786

Mark Brandenburg:
I.R. 1 founded 1615, garrison: Berlin cadre: Brandenburg, draft: Brandenburg
I.R. 6 founded 1675, garrison: Potsdam cadre: Brandenburg, draft: Army, Silesia
I.R. 12 founded 1685, garrison: Prenzlau cadre: Brandenburg, draft: Brandenburg
I.R. 13 founded 1686, garrison: Berlin cadre: France, draft: Brandenburg
I.R. 15 founded 1688, garrison: Potsdam cadre: Brandenburg, draft: Brandenburg
I.R. 18 founded 1698, garrison: Potsdam cadre: Brandenburg, draft: Brandenburg
I.R. 19 founded 1702, garrison: Berlin cadre: Brandenburg, draft: Brandenburg
I.R. 23 founded 1714, garrison: Berlin cadre: Brandenburg, draft: Brandenburg
I.R. 24 founded 1715, garrison: Frankfurt/Oder cadre: Brandenburg, draft: Brandenburg
I.R. 25 founded 1715, garrison: Berlin cadre: Brandenburg, draft: Brandenburg
I.R. 26 founded 1714, garrison: Berlin cadre: East Prussia, draft: Brandenburg
I.R. 27 founded 1715, garrison: Stendal cadre: Pomerania, draft: Brandenburg
I.R. 34 founded 1740, garrison: Ruppin cadre: Brandenburg, East Prussia, draft: Brandenburg
I.R. 35 founded 1740, garrison: Spandau cadre: Brandenburg, East Prussia, draft: Silesia, recruits
I.R. 36 founded 1740, garrison: Brandenburg cadre: Brandenburg, Pomerania, draft: Pomerania
I.R. 39 founded 1740, garrison: Königsberg/N. cadre: Braunschweig, draft: Braunschweig
I.R. 46 founded 1743, garrison: Berlin cadre: Württemberg, Brandenburg, draft: Brandenburg
G.R. VII founded 1741, garrison: Eberswalde etc.

Pomerania
I.R. 7 founded 1676, garrison: Stettin cadre: Brandenburg, draft: Pomerania
I.R. 8 founded 1679, garrison: Stettin cadre: Brandenburg, draft: Pomerania
I.R. 17 founded 1693, garrison: Köslin cadre: Brandenburg, draft: Pomerania
I.R. 22 founded 1713, garrison: Stargard cadre: Brandenburg, draft: Pomerania
I.R. 30 founded 1728, garrison: Anklam cadre: Brandenburg, draft: Pomerania
G.R. III founded 1718, garrison: Kolberg etc.

East Prussia
I.R. 2 founded 1655, garrison: Königsberg cadre: Brandenburg, Pomerania, draft: Prussia
I.R. 4 founded 1672, garrison: Elbing cadre: Brandenburg, draft: Prussia
I.R. 11 founded 1685, garrison: Königsberg cadre: Rhineland, draft: Prussia
I.R. 14 founded 1626, garrison: Bartenstein cadre: Prussia, draft: Prussia
I.R. 16 founded 1689, garrison: Königsberg cadre: Westphalia, Prussia, draft: Prussia

West Prussia
I.R. 51 founded 1773, garrison: Marienburg cadre: Prussia, draft: Prussia
I.R. 52 founded 1773, garrison: Preussisch Holland cadre: Prussia, draft: Prussia
I.R. 53 founded 1773, garrison: Braunsberg cadre: Prussia, draft: Prussia
I.R. 54 founded 1773, garrison: Graudenz cadre: Prussia, draft: Prussia
I.R. 55 founded 1774, garrison: Preussisch Stargard cadre: Prussia, draft: Prussia
G.R. I founded 1718, garrison: Rastenburg etc.
G.R. II founded 1717, garrison: Danzig etc.
G.R. XI founded 1744, garrison: Allenstein etc.

Silesia
I.R. 28 founded 1723, garrison: Brieg cadre: Rhineland, draft: Silesia
I.R. 29 founded 1723, garrison: Breslau cadre: Rhineland, draft: Silesia
I.R. 31 founded 1729, garrison: Breslau cadre: Rhineland, draft: Silesia
I.R. 32 founded 1743, garrison: Neisse cadre: Westphalia, Rhineland, draft: Silesia
I.R. 33 founded 1740, garrison: Glatz cadre: Magdeburg, draft: Silesia
I.R. 37 founded 1740, garrison: Glogau cadre: Brandenburg, draft: Silesia
I.R. 38 founded 1740, garrison: Frankenstein cadre: Prussia, draft: Silesia
I.R. 40 founded 1740, garrison: Schweidnitz cadre: Thuringia, draft: Silesia
I.R. 42 founded 1741, garrison: Breslau cadre: Silesia, draft: Silesia
I.R. 43 founded 1741, garrison: Liegnitz cadre: Silesia, draft: Silesia
I.R. 49 founded 1742, garrison: Neisse cadre: Silesia, draft: Silesia
I.R. 50 founded 1773, garrison: Silberberg cadre: Switzerland, Württemberg, draft: Silesia
G.R. V founded 1741, garrison: Jauer, Striegau etc.
G.R. VI founded 1740, garrison: Cosel, Namslau
G.R. VIII founded 1742, garrison: Glatz, Reichenbach
G.R. X founded 1742, garrison: Neisse, Ottmachau

Magdeburg/Halberstadt
I.R. 3 founded 1665, garrison: Halle cadre: Bavaria, Hesse, draft: Magdeburg
I.R. 5 founded 1655, garrison: Megdeburg cadre: Prussia, draft: Magdeburg
I.R. 20 founded 1707, garrison: Magdeburg cadre: Magdeburg, draft: Magdeburg
I.R. 21 founded 1713, garrison: Halberstadt cadre: Westphalia, Pomerania, draft: Magdeburg
I.R. 47 founded 1743, garrison: Burg cadre: Holstein, recruits, draft: Upper Silesia
G.R. IV founded 1736, garrison: Aken, Könnern, Löbejün

Cleves, Mark, Revensberg, Minden, Ostfriesland
I.R. 9 founded 1680, garrison: Hamm, Soest cadre: Westphalia, draft: Westphalia
I.R. 10 founded 1683, garrison: Bielefeld cadre: Prussia, draft: Westphalia
I.R. 41 founded 1741, garrison: Minden cadre: Württemberg, draft: Westphalia
I.R. 44 founded 1742, garrison: Wesel cadre: Württemberg, draft: Army
I.R. 45 founded 1743, garrison: Wesel cadre: Westphalia, draft: Army
I.R. 48 founded 1756, garrison: Wesel cadre: Rhineland, Westphalia, draft: Army
G.R. IX founded 1743, garrison: Geldern
G.R. XII founded 1744, garrison: Emden

The foundings before 1740 and after that date are differentiated clearly. Of note are the founding places, their geographical distribution and the troops that formed their cadre. The changing garrisons are stated here as of 1786. The notation "draft: Army" means that the regiment had no drafting area but took replacements from the army in general.

Soldiers from the same region, often well acquainted or related,defined the atmosphere, togetherness and readiness to perform of the regiment. They trusted the King and were loyal to him. The old provinces provided the original troops, and the newer ones strove to follow their example.

GARRISON REGIMENT NO. 1

Commanders of the Regiment

1718	10/28	Colonel Carl de Brion
1728	5/3	Colonel Ludwig de l'Hospital
1755	4/20	Colonel Christoph Georg von Luck
1757	3/4	Colonel Werner Friedrich von Puttkamer
1772	1/2	Colonel Friedrich Sylvius von Hallmann

After his accession on February 23, 1713 King Friedrich Wilhelm I ordered all garrison troop units to be used to establish new field regiments, as until then only a difference in function had existed between the two. Their grenadiers in any case belonged to the field army, as also under Frederick the Great, on an equal basis and proved their worth as so-called "standing battalions", as for example those of Magdeburg and Königsberg. After the Pomeranian campaign, from 1716 on, he gathered soldiers who were no longer ranked, whether because of short stature, age or invalidism, for sentry duty or occupying units, and founded five new garrison battalions in 1717 and 1718; step by step they grew to regimental strength from 1721 on. Since then the precision of the terms "regiment" and "battalion" has varied. In 1725 a sixth one was added, and two were transferred to I.R. 32 and 30. In 1740 there were still four, two in East Prussia, one in Far Pomerania and one in Magdeburg, focal points of defense at that time. Their main job was to protect the fortresses in time of war. The officers were often townsmen, older men with war records, or had earned the right to rise from the ranks of junior officers; regimental commanders, on the other hand, could alternate between field regiments and garrison troops. The soldiers, though, received lower pay, less fancy uniforms, and from Garrison Regiment 3 on, one-color flags without shiny surfaces. They lacked prestige and honors and performed only monotonous service. Many regiments—such as Garrison Regiment VII—served as repositories for soldiers transferred as punishment, which did not improve their reputation. Still in all, the soldiers were capable, and men do develop by having demands made on them. Friedrich II took over the four units and made immediate use of the Magdeburg Garrison Regiment of Persod to establish Fusilier Regiment No. 33, but turned the disestablished Guard Regiment No. 6 into the new Weyher Garrison Regiment No. 4. Beginning with No. 5, all garrison units were originated by the King. According to Bleckwenn, the obvious decline in their social status came only after 1740, perhaps as a result of the army's expansion and wartime fame. The King called a third to half of the garrison troops into field service, aside from new establishments, particularly because of the increasing losses in the Seven Years' War. After the King's death, 24 'green' fusilier battalions were formed as new light infantry units in 1787. In 1788 all garrison regiments were disbanded and their men used to form depot battalions in the field regiments, which became the Third Musketeer Battalion in 1794. Since 1794 there was a separate invalid company in every regiment. These Third Battalions proved themselves well in 1806-1807, especially in defending the eastern fortresses; most of them remained in existence and in 1808, as tested human potential, decisive in catastrophe, they formed the cadres for the new regiments.

Garrison Regiment I was founded in 1718 in Memel, originally with two and until 1721 with five companies, under Colonelde Brion, who took it over on October 28 and served as its chief for ten years. His successor, de l'Hospital, led it for 27 years. As ordered on November 29, 1743 the battalion was to be brought up to full two-battalion regimental strength as of February 1, 1744 by dividing the companies and recruiting in the Empire and Poland. By strengthening the garrison troops, which began in December of 1756, the King enlarged the regiment to four battalions, but after peace was made in 1763 the conditions of September 1, 1756 were restablished, with only two battalions. It grew to four battalions again, obviously regarded as "wartime strength", in 1772. In the Second Silesian War the grenadier companies proved themselves for the first time at Hohenfriedeberg, so well, in fact, that they were always relied on in the future. From 1756 to 1763 they were fighting, at Gross-Jägersdorf in 1757, Zorndorf in 1758, Kay and Kunersdorf in 1759 and Torgau in 1760, and these are just three victories and four battles with the Russians, a sign of stability. The whole regiment took part in the Succession War of 1778. Its garrison was Memel from 1718 to 1745, interrupted by a stay in Stettin in 1734-1735. From 1746 to 1756 it was in Känigsberg, Gumbinnen, Memel and Pillau, and in Stettin again briefly in 1760. From 1764 to 1782 it was in Memel and Gumbinnen again with two, then four battalions, from 1783 to 1788 in Rastenburg, Gumbinnen, Rössel and Angerburg. In 1788 it formed the depot battalions of I.R. 11, 14, 16, 30 and 39 in East Prussia, Pomerania and the Mark. One company secured Fort Lyck as its occupation force until 1792. Its recruits came from the East Prussian cities of Memel, Wehlau, Tapiau, Hohenstein, Nordenburg, Schirwindt and Stallupönen from 1733 to 1788.

Garnison-Regiment
Nr. I

GARRISON REGIMENT NO. II

Commanders of the Regiment

1717	Feb.	Colonel Georg Bogislav von Wobeser
1723	1/11	Colonel Caspar Dietlof von Winterfeld
1725	2/3	Colonel Peter de Sers
1727	12/28	Colonel Johann de Natalis
1742	1/30	Lieutenant Colonel Georg Werner von der Schulenburg
1743	8/28	Colonel Christoph Ernst von Roeder
1754	8/5	Colonel Hans Sigismund von Sydow
1773	5/18	Colonel Georg Wolf von Tümpling
1777	9/9	Colonel Georg Lorenz von Pirch

The battalion established in Pillau as of March 1717 under Colonel Georg Bogislav von Wobeser developed into the King's best garrison regiment. The number of its soldiers mustered out of field regiments was increased by recruiting to form five companies, funded, as were all battalions—except the one in Geldern—from the invalid budget. In 1724 a new transfer of mustered-out soldiers created a sixth company. On June 1, 1739 the four existing garrison battalions were put on the budget of the general war funds; in terms of treatment, this made a "social establishment" into a troop. Naturally, this raised the internal question of competition with the comparable effective use of the field regiments. As ordered on January 20, 1744, the battalion divided its companies as of that February 1 and filled them with men recruited from the Empire and Poland. At the same time, it gave one of its grenadier companies to the new Garrison Regiment XI in Angerburg and made the remaining grenadier company into two. Through this reorganization in 1744 the King gained fourteen battalions in all for his infantry. On July 27, 1756 Field Marshal von Lehwaldt in East Prussia was ordered to strengthen the regiment to four battalions through transfers from the field regiments there and draftees. This doubling prepared it for mobilization. As ordered on August 7, the founding of the two battalions took place in Lochstädt. I.R. 11 and 16 provided 310 and 330 men respectively, Dragoon Regiments 6, 7 and 8 gave 310, 200 and 200 men each, making 1350 in all from five field regiments. Fourteen garrison battalions were created in all, twelve of which were established only upon mobilization. Garrisons were Pillau from 1717 to 1733, Stettin in 1734-1735, Pillau from 1736 to 1738, Königsberg and Pillau in 1739, Pillau from 1740 to 1742, Königsberg and Pillau from 1743 to 1745, Treuenbrietzen and Berlin in 1745. The transfers out of Pillau took place in connection with preparations for war or actual war conditions. The grenadier companies fought successfully at Hohenfriedeberg in 1745. Its recruits came from the cities of Pillau, Mohrungen, Saalfeld, Lyck, Marienwerder, Rosenberg and Johannisburg. The peacetime years from 1746 to 1755 saw it in Pillau, Fischhausen and Friedrichsburg.

In the field army in 1757, it fought in the second line at Gross-Jägersdorf on August 30 and in the thick Norkitt Forest, amid powder smoke and battle noise, it began to shoot at its own first line by mistake, taking them for the Russians. This confusion made the Russian counterattack possible. In 1758 and 1758 it besieged Schweidnitz, secured Silesia and Breslau. At Zorndorf in 1758 only the grenadiers saw action, likewise at Kay and Kunersdorf in 1759. From 1760 on the First Battalion belonged to the army of Prince Heinrich in Silesia and Saxony. The other battalions were at Breslau and Neisse at first, in Neisse, Trachenberg and Militsch in 1762. The First Battalion and grenadiers fought in the Battle of Torgau on November 3, 1760, as also at Freiberg on August 29, 1762 in the brigade of Colonel von Taube, Forcade Column. While all other garrison battalions were returnd to their strength of September 1, 1756 after peace was made in 1763, G.R. II retained three battalions because of the disbanding of the New Garrison Regiment; they were divided among Pillau (seven companies), Fischhausen (three), Barten (two), Tapiau (two) and Domnau (six). The Fourth Battalion was reestablished in 1772 in the course of the Polish troubles. From 1783 to 1788 ten companies were in Danzig, four in Zinten, three each in Preussisch Eylau and Kreuzburg/Pr. Although its uniform was made more impressive by the addition of white cuffs, the regimental chief made improvements in 1778, and in 1788 it formed depot battalions for I.R. 2 and 4 in East Prussia, I.R. 8, 17 and 22 in Pomerania, and I.R. 52 in West Prussia; these later became musketeers.

Garnison-Regiment Nr. II

GARRISON REGIMENT NO. III

Commanders of the Regiment

1718	8/9	Colonel Sigismund von Sack, later Major General
1740	9/7	Lieutenant General Georg Rudolf von Glaubitz, former Commander of I.R. No. 4
1740	10/10	Colonel Volrath Hellermann, ennobled 1743
1756	12/21	Colonel Georg Arnold von Grolman
1763	5/2	Colonel August Friedrich, Marshal von Biberstein
1767	4/27	Colonel Ernst Friedrich von der Heyden, former Commander of the Berlin State Regiment
1772	6/1	Colonel Detlof, Baron von Vietinghoff

The Colberg Garrison Regiment was created in August of 1718, with three companies. In 1721 it was strengthened to five companies, keeping it in line with the Peenemünde Company founded in 1720. In addition to Colberg itself, the Colberg Battalion also secured the Peenemünde battlements, the Anklam ferry and Swinemünde, thus being responsible for safeguarding both sides of the mouth of the Oder, more than 135 kilometers as the crow flies. This was scarcely an exaggeration in a time when the Swedes still owned Near Pomerania! Since August 9, 1718 it chief was Colonel von Sack, who served as such for 22 years and was promoted to Major General in 1740. Its 1734-1735 transfer to Magdeburg was a preparatory move for mobilization because of the War of the Polish Succession and the pressure on Berg in the event that the Electors Palatine should die out. In 1740 it supplied a large part of the manpower for the founding of I.R. 37, plus its grenadiers for I.R. 36, which were urgently needed. Refilling took place gradually. In 1741 the battalion reinforced the protection of Stettin by I.R. 39. There the organizing of a new garrison regiment had been in progress since February of that year; it was later to become G.R. VII, which reached full strength that autumn. Thus the battalion was relieved of the responsibility for the mouth of the Oder. This may also explain why its enlargement to two battalions took place only by order of the King in December of 1756. Next it was sent with G.R. IV to occupy the Saxon forts on the Elbe. Its garrison from 1718 to 1742 was Colberg, but this was interrupted by its use in Magdeburg in 1734-1735 and Stettin in 1741-1742. After protecting Stettin in 1744 it joined the Berlin occupation in 1745, under Major General Count Hacke, when Berlin was threatened by Saxony—especially by Grünne's Corps—before the victories at Hohenfriedeberg and Kesselsdorf eased the situation. From 1746 on the grenadiers were in Treuenbrietzen. In the peacetime years from 1747 to 1755 it was back in Colberg, as also after the Seven Years' War and the Succession War of 1778. Its recruits came from the city of Colberg and the smaller towns of Bublitz, Bütow and Usedom almost without change from 1733 to 1788.

While the grenadiers were fighting at Prague, Kolin, Schweidnitz and Leuthen in 1757, Hochkirch in 1758 and Torgau in 1760, the Second Battalion was on the field budget in 1717. In 1758 and 1759 the regiment was part of the occupation troop for Prince Heinrich's Saxon Corps and provided protection to cities and freeing the active regiments for field service, the First Battalion being in Torgau, the Second in Dresden. On September 27, 1760 the First Battalion was taken prisoner in Torgau, as was the Second in Wittenberg on October 14. Because of the interrupted exchange of prisoners, it could not be reconstituted at that time. Only when the war ended in 1763 was it reformed in Colberg and raised to the strength of a battalion by adding to its returning prisoners the Free Corps of Major Carl Ludwig von Schony, a Hungarian in the service of Württemberg, with sixteen officers, 36 junior officers and 600 men. It remained in Colberg until 1788. When it was disbanded it provided the depot battalions, later third battalions, for I.R. 1 and 19 in Berlin.

Garnison-Regiment
Nr. III

GARRISON REGIMENT NO. IV

Commanders of the Regiment

1736	10/1	Colonel Caspar Jochen von Wachholtz
1737	1/29	Colonel Peter d'Arbaud
1739	10/15	Colonel Andreas de Persod, later Commander of I.R. 33
1740	6/23	Major General Adam von Weyher, died 12/10/1744
1740-	1745	vacant
1746	1/4	Colonel Jakob Heinrich von Grape
1759	1/7	Major General Martin Eberhard von Jungkenn Müntzer von Mohrenstamm, former Commander of I.R. 44
1760	2/5	Colonel Ewald Georg von Lettow
1763	2/16	Lieutenant Colonel Friedrich Franz Ernst, Baron von Plotho
1766	3/24	Colonel Franz Thuro von Groscreutz
1769	7/3	Colonel Ludwig von Gohr
1772	5/23	Colonel Peter Georg von Puttkamer
1775	9/5	Colonel Heinrich Ludwig von Gotter
1782	9/8	Colonel Valentin Friedrich von Rüchel
1784	9/27	Major General Otto Casimir von Meerscheidt, known as Hüllesem

Out of the Imperial contingent of the four Princes of Anhalt from Dessau, Zerbst, Bernburg and Köthen, which had been stationed on the upper Rhine in 1734-1735, a garrison battalion was formed in Aken under Colonel von Wachholtz after the troops had been taken into Prussian service. Since at first it had only four companies, it received transfers from the disbanded "New Garrison" of Magdeburg and I.R. 3 of Anhalt-Dessau to enlarge it to one grenadier and five musketeer companies. In 1737 it was transferred to Magdeburg. When its commander, Colonel de Persod, used it to form the new Fusilier Regiment 33 in 1740, a new garrison battalion was created simultaneously to take its place in Magdeburg, composed of men no longer fit for field service and "those of bad conduct, chiefly foreigners" from I.R. 6, formerly the King's Guard Regiment and now reduced to one battalion. The old and new battalions had only their location in common. Colonel von Weyher was the commander. In this wave of foundation the King created eighteen new infantry battalions immediately after his accession. In 1741 it stayed in Magdeburg along with Fusilier Regiments 35 and 40. As of 1742 the grenadiers were paid as much as field soldiers. In 1742 the two companies of G.R. IV were ordered to Treuenbrietzen as of August 1. In 1748 the battalion was transferred to Aken and Könnern/Saale, in 1749 Löbejün was added. The regiment never had a draft area; from 1736 to 1740 its replacements came from the areas of Seehausen and Alsleben, from 1740 to 1788, interestingly enough, from I.R. 3 in Halle, I.R. 5and 20 in Magdeburg, I.R. 21 in Halberstadt, I.R. 27 in Stendal and I.R. 47 in Burg, thus from central German field regiments. In December of 1756 it was ordered to double its strength to two battalions as of March 1757, as it was intended to be used, as were Garrison Regiments 3 and 7, for reinforce the securing of Saxon forts along the Elbe. In the figures of the King's field budget at the end of 1756 the new battalion is already included. In 1759 it received the former chief of Fusilier Regiment 44, Major General von Jungkenn, known as Müntzer von Mohrenstamm, as its new chief, after it had gone with him as an occupation troop in Prince Heinrich's Saxon Corps. It was located principally in Wittenberg and Torgau. In 1759 both battalions defended Wittenberg under Major General von Horn who, out of lack of determination, surrendered the fortress to the Imperial army on August 21 in exchange for free departure. When Major General von Wunsch turned up immediately thereafter with his strengthened free regiment, bringing along parts of the arriving Kleist Corps as well, he took command of the regiment that was just leaving and the occupying forces at Torgau and took Wittenberg right back again without opposition, as well as Torgau in a night attack, turning then toward Dresden. In the autumn of 1760 Hülsen had to defend Saxony alone against fourfold superiority and retreat across the Elbe, whose bridges at Torgau were set afire by gunfire. His right wing was supposed to get support from the fortress. But one day later, on September 27, Major von Normann surrendered it to the enemy. The somewhat weakened regiment went into captivity with the occupying troops, and there was no more exchange of prisoners until the war ended. In 1761 and 1762 it was not reformed. On the other hand, the number of free battalions increased considerably, whereby the Courbiere, expanded to a regiment by Major Charies Guischardt, known as Quintus Icilius, by means of recruiting and with the Second Battalion, took its place. After the prisoners returned in 1763 the garrison battalion was organized anew in Aken, Könnern and Löbejün under Lieutenant Colonel Baron von Plotho. To fill it, the King transferred the second battalion of the experienced Courbiere Free Regiment to it. In 1788 it was disbanded to make up the depot battalions, later third battalions, of I.R. 9 and 10 in Hamm and Bielefeld, which capitulated ingloriously in Hamelin and Nienburg in 1806.

Garnison-Regiment
Nr. IV

GARRISON REGIMENT NO. V

Commanders of the Regiment

1741	8/19	Major General Christopf Friedrich von Thmen
1743	8/6	Colonel Friedrich Julius von Mützschefahl, later Major General
1759	2/18	Colonel Gustav Adolf von Sydow, later Major General
1763	9/10	Colonel Christian Friedrich von Berner
1770	7/31	Colonel Johann Benjamin von Hasslocher
1771	7/31	Colonel Adam Friedrich von Arnstädt
1778	4/5	Colonel Paul von Natalis

The establishment of a garrison regiment for the Fortress of Glogau, just captured on March 9, 1741, began in August of that year, after the Battle of Mollwitz and proceeded slowly. The formation was directed by Colonel von Schwerin of I.R. 36, who was then the Commandant of Glogau, and the command was given to Major General von Thümen, who had resigned command of Dragoon Regiment No. 7, on August 19, 1741. Early in February of 1742 the regiment still did not have guns, although recruitment had been going on for months. Only at year's end was the regiment complete, with ten musketeer and two grenadier companies, and at first named after the place of formation or the intended fort. Three companies of the regiment were created at the end of 1741 of invalids of the regiments in Silesia, capable of garrison service. By order of the King on July 15, 1742 they were exchanged with the Küstrin and Peitz garrison companies, while the officers stayed where they were. The invalids were paid the higher field wages in full, as the King promised. On August 6, 1743 Colonel von Mützschefahl became the regimental chief, and was promoted to Major General in 1756. Glogau was not only the entryway to Lower Silesia, but important in securing the Oder and the crossings to Posen. The garrison was Glogau, plus Crossen, Reppen, Züllichau, Beeskow and Sommerfeld as of 1755,as well as Fürstenwalde from 1747 to 1750. As of 1764 it had one battalion each in Crossen, Züllichau, Jauer and Striegau, as well as Neumarkt in Middle Silesia. Until 1747 its draft area was the region around Glogau. After that replacements came from I.R. 29 in Breslau, I.R. 37 in Glogau and I.R. 38 in Liegnitz, then Breslau and Frankenstein. The grenadiers were always in Glogau and formed the permanent Grenadier Battalion 5 along with those of G.R. X in Breslau. In view of the threat of war in 1749, the King wanted to free field regiments of occupation duty in the Silesian forts; on August 13, 1753 he requested additional funds for barracks for ten companies in Glogau. On July 8, 1754 he gave instructions for the formation of two additional battalions, but without grenadiers. On January 19, 1755 the Silesian regiments were to report how many men of normal height they had, "but who are vagabonds and of whom the regiment is not really sure". Thus 589 men were assembled, while other regiments provided 314 men and the regiment itself supplied a hundred. Along with them came 447 Silesian recruits. These 1350 men formed the Third and Fourth Battalions under the budget of June 1, 1755, established as of August 15 of that year in Jauer, Striegau and Neumarkt. The shortening of training by ten weeks covered the cost of uniforms and equipment. The regiment was intended to occupy Schweidnitz in wartime, and it was kept at full strength after 1763.

In 1756 and 1757 the grenadiers belonged to Schwerin's Corps, which secured Silesia, and then helped occupy Cosel. On November 12, 1757 three battalions of the regiment were captured in Schweidnitz. In 1758 the First and Second Battalions belonged to the King's Army, which marched to Olmütz after recapturing Schweidnitz. On June 28 of that year, while guarding the great ammunition and cash transport train at Domstadtl under Lieutenant Colonel von Mosel, they suffered the well-known attack that ended disastrously on June 30. As a rear unit, they could go back to Troppau. In 1759 they stayed with the King's Army, which had summoned both battalions from nearby Glogau to strengthen Wedell's Corps. On July 23 of that year they took part in the four stormings of the Palzig Heights, which were held by the Russians; despite losing over 30% of their men, they were defeated. After that they were not spared the Battle of Kunersdorf on August 12. Except for the grenadiers, they were the only garrison troops that were called in. In 1760 both battalions were on the field budget to replace five field regiments, but instead joined Prince Heinrich's Corps, then the Saxon Corps in 1761, which cleverly held the narrow area between the Elbe, Triebisch and Freiberg Mulde. In 1762 the King had both battalions strengthened by 300 men ("new foot") and equipped like a field regiment. They stayed on in Saxony and had practically become parts of field troops. The Third and Fourth Battalions had served as occupation troops in Silesia. In 1788 the regiment provided the depot battalions for I.R. 3 in Halle, I.R. 5 and 20 in Magdeburg, I.R. 21 in Halberstadt, I.R. 27 in Stendal, I.R. 36 in Brandenburg and I.R. 41 in Minden, thus in central Germany; under various captains, they surrendered as third battalions in 1806.

Garnison-Regiment
Nr. V

GARRISON REGIMENT NO. VI

Commanders of the Regiment

1741	7/15	Colonel Caspar Heinrich von Stechow
1741	8/21	Major General Heinrich von Saldern
1745	10/30	Major General Johann Georg von Lehmann
1750	12/15	Lieutenant General Heinrich Günther Gottfried von Bosse
1753	12/10	Major General Christian Friedrich von Lattorf
1762	4/4	Colonel Gerhard Alexander, Baron von Sass

After the Battle of Mollwitz on April 10, 1741 and the taking of Brieg on May 4 of that year by General von Linger and Colonel von Walrave, 400 men from the occupation force—"some willingly, some forced"—entered Prussian service. They became the core of the "Brieg Garrison Regiment", whose founding began as early as that June. The establishment was directed by Colonel von Wurmb as its commander, originally an infantryman but later a hussar. The chief's position was given to Colonel von Stechow, former commander of I.R. 29, in July, by which time five companies with 460 men had been set up, to be strengthened by 800 "whitecoat" prisoners released by the Austrians at Grottkau on July 9. These "whitecoats", half-military fort construction workers and baggage-train personnel, had first been used in 1735, and wore white linen clothing instead of uniforms. A regiment had up to eighty of them. The size of the regiment was set at two batalions. Instead of the two grenadier companies of East Prussian garrison battalions, it received those of I.R. 41, which had been taken over from Württemberg. In the winter of 1741-1742 it had to give up the greater part of its manpower for the founding of I.R. 42, receiving in return recruits from Upper Silesia and Bohemia, which Hereditary Prince Leopold "had to have assembled, whether by kindness or by force". Thus it reached its full strength only at the end of 1742. In mid-April 1745 the First Battalion joined Hautcharmoy's small corps, which had the job of protecting northern Upper Silesia from Hungarian attacks. If war action came close to the sites of garrison regiments, they were to be called in. As the occupation of the as yet unfinished Fortress of Cosel, it was taken prisoner on May 26 after the death of its chief and commandant, Major General von Saldern. As of May 1753 its grenadiers formed the Grenadier Battalion No. 6 of Plötz. As with Garrison Regiment No. V, the King, in view of the threat of war, ordered the regiment expanded at four battalions; ordered on June 19, 1756, this took place quickly, starting on August 1 of that year, solely by dividing the ten companies at hand. Recruits filled them up again. The Third and Fourth Battalions came to Cosel as occupation troops. After the peace of 1763 these battalions also remained in existence, as with all Silesian garrison regiments. From 1741 to 1746 the regiment was stationed only in Brieg, from 1747 to 1756 in the fortresses of Cosel and Brieg and in Namslau, 1751-1752 only in Cosel for a time. Replacements were recruited in Upper Silesia and Bohemia from 1741 to 1743. From 1743 to 1747 they came from the Brieg area, from 1747 to 1788 from I.R. 29 and 31 in Breslau, I.R. 42 in Reichenbach and I.R. 47 in Burg, along with Upper Silesian recruits. In the summer of 1756 the forts were put on a defensive basis, the cannons brought in, palisades set up and men on furlough called back two weeks earlier.

In 1758 the KIng took the Fourth Battalion along to the campaign in northern Moravia, which was to depend on the capture of Olmütz, but ended in early June with the retreat to Königgrätz. In 1759 it was again in the King's Army when needed, sometimes as a field troop. The hesitant use of garrison troops surely had less to do with their quality than with their main use in defending forts and homeland, plus the fact that, without draft areas of their own, they could hardly take replacements from the already weakened field regiments.

In 1760 it joined six other garrison regiments to replace five weakened field regiments on the field budget and belonged to Prince Heinrich's Corps. The human potential of the Prussian infantry was not inexhaustible. In 1761 the Fourth Battalion belonged to Goltz's Corps, which was supposed to advance on Posen; as of June 30 it was taken over by Zieten. Simply in terms of expense, the intention was impossible, and Zieten marched past Breslau to camp at Bunzelwitz. In 1762 it was with the army in Silesia but attracted no particular attention. But an army in the field also needs forces to the rear, for marching, protection of supplying, securing service, requisitioning, outpost service, flank protection and small engagements.

From 1764 to 1768 the regiment had three battalions in Cosel and one in Namslau, uninterrupted by the Succession War in 1788. In that year it supplied the depot battalions for I.R. 28 in Brieg, I.R. 29 and 31 in Breslau, I.R. 38 in Frankenstein and I.R. 43 in Liegnitz; these later became third battalions and still later joined the Grenadier Regiment King Friedrich Wilhelm Ii (First Silesian) No. 10 in Schweidnitz and Grenadier Regiment King Friedrich III (Second Silesian) No. 11 in Breslau.

Garnison-Regiment
Nr. VI

GARRISON REGIMENT NO. VII

Commanders of the Regiment

1741	10/31	Major General Carl Wilhelm von Bredow
1746	3/10	Colonel Hans Christoph von Jeetze
1754	4/13	Colonel Christian Henning von Lange
1760	2/20	Major General Joachim Christian Friedrich von Itzenplitz
1766	4/20	Major General Friedrich von Rentzel, later of I.R. 23
1766	5/31	Major General Christian Ernst von Puttkamer
1771	2/2	Colonel Georg Lorenz von Kowalsky

In February of 1741 the establishment of the "Stettin Garrison Regiment" began in Stettin with the transfer of two battalions from the Grenadier Guard Battalion No. 6 of Potsdam, the men being recruits sent back because they were too short or not suitable for field duty. Colonel von Kreytzen of I.R. 29 directed the formation and had the duties of the commandant. The first Austrian war prisoners who came into Prussian service, plus several hundred volunteers from the "New Garrison Regiment" of Stettin filled the companies. The two grenadier companies were made up of Prince Schwarzburg's Company of Captain von Spitznass, taken over on May 26 in Demmin, with three officers and 83 men. They had been stationed in Rostock since 1736, intended for the Imperial action against Magdeburg. In August of that year the regiment was almost at full strength. On October 31 the King turned it over to Major General von Bredow, appointed as Commandant of Stettin, who had turned his I.R. 7 over to the Prince of Braunschweig-Bevern. In 1744 the "New Garrisons" were not assembled. Instead the regiment protected Stettin, which guarded the mouth of the Oder, the Uckermark and connections to Far Pomerania. When the situation on the Saxon border became critical after Soor in 1745 and Berlin had to be protected, it reinforced the weak occupying troops in Berlin under Major General Count Hacke, consisting only of garrison troops, recruits and armed citizens. The victory at Kesselsdorf removed the danger. In 1756 the regiment was enlarged by another two battalions by the budget of June 1; they were formed as of August 15. Again the King called on a regiment of the Prince of Schwarzburg, which he had requested since its return from The Netherlands in 1749. On April 8, 1756 it could be taken over, comprising eighteen officers, 32 junior officers, twelve musicians and 299 privates. In July they were joined by 300 draftees and 200 recruited 'foreigners' from the ten Magdeburg infantry and cavalry regiments plus 300 draftees of the Westphalian I.R. 9, 10 and 41, a total of 1161 men. These battalions had no grenadiers' they were set up in Müncheberg and Strausberg, east of Berlin. After establishment in Stettin, garrisons were Neustadt-Eberswalde, Bernau, Templin, Oderberg and Angermünde from 1741-1742 to 1756, as well as Liebenwalde for a time in 1753-1754. Before the Seven Years' War began, the First Battalion went to Glogau, the Second to Breslau, the Third and Fourth to Berlin. Until 1788 its replacements came from Regiments 1, 13, 19, 23, 25, 26 and 46 in Berlin. At the beginning of 1757 the King had the two Berlin battalions doubled; the Fifth was formed that June in Dresden, the Sixth in September in Berlin. The Third and Fifth Battalions were assigned to occupy Saxon forts on the Elbe and were included in the tally of manpower.

When on October 16 a weak raiding corps under Hadik reached the Berlin suburb of Köpenick, site of the Kings headquarters for military technology, the amazed commandant, General von Rochow, sent six companies of the regiment under Major von Tesmar against five times their strength. After their commander's death and a brave fight, they were defeated. The enemy withdrew the next day. Things had gone well again. After experiencing defeat on the Lohe on November 22 of that year, the Second Battalion faced the surrender of Breslau three days later, with free departure under panic conditions. In 1758 the Third and Fifth Battalions were in the Saxon Corps, while at the end of that July the Sixth Battalion, under Lieutenant General von Schorlemmer, had to strengthen the occupation of Küstrin against the Russians. In 1759 the Third and Fifth Battalions stayed with the Saxon Corpsas an occupation troop. The Elbe forts were very important and had required great numbers of field units. The two battalions were included in the tragic surrender of Dresden on September 4 under Count Schmettau. In 1760 they defended Leipzig and had to march to Magdeburg at Hülsen's command. The Berlin battalions of the regiment were captured by the Russians on October 9, returning only in 1762. In 1761 the battalions could not be reconstituted. After the Battle of Burkersdorf in 1762, the First Battalion was summoned from Glogau to help besiege Schweidnitz under General von Tauentzien. All the battalions were on field service at times.

From 1764 to 1788 the regiment's four battalions were stationed in Neustadt-Eberswalde, Angermünde, Templin, Küstrin, Bernau and Spandau, and from 1764 to 1767 also in Mohrin, from 1764 to 1773 in Reppen and from 1767 to 1771 in Strausberg, all locations around Berlin to the north and east. When it was disbanded in 1788, three companies each went to I.R. 13 in Berlin, I.R. 128 in Potsdam, I.R. 23, 25, 26 and 46 in Berlin, and one company each to I.R. 19 in Berlin and I.R. 36 in Brandenburg/Havel to serve as depot battalions, later as third battalions. They ceased to exist in 1806 except for the Third Battalion of I.R. 46 in East Prussia, which passed via the First Silesian Infantry Regiment to the Grenadier Regiment King Friedrich Wilhelm II (First Silesian) No. 10 of Schweidnitz in 1808.

Garnison-Regiment
Nr. VII

GARRISON REGIMENT NO. VIII

Commanders of the Regiment

1741	9/5	Colonel Carl, Baron von der Reck
1745	10/30	Major General Rudolf Curt Leberecht, Baron von Loeben
1746	11/30	Colonel Caspar Friedrich von Knobelsdorff
1748	11/25	Colonel Georg Ernst von Nettelhorst
1757	6/10	Colonel Friedrich, Baron Quadt von Wickeradt
1763	2/11	Colonel Franziskus de le Noble
1772	2/23	Colonel Carl Benedikt von Bremer
1778	11/7	Colonel Sigismund August von Berrenhauer
1782	1/5	Colonel Ulrich von Heuking

In the winter of 1741-1742 the King decided that a garrison regiment for Neisse should be founded. On September 5, 1741 Colonel Baron von der Reck was given permission to set up a battalion of Imperial recruits, who were to be recruited and assembled in Minden and brought to the Mark Brandenburg. In Müncheberg, where the formation began, only 370 men were assembled as of February 1742. By order of March 30 of that year, Garrison Regiment VII in Stettin supplied 500 men to form the Second Battalion, receiving in exchange 500 recruits from Upper Silesia and Moravia. In the summer of that year the First Battalion marched to Silesia, while the Second followed in 1743. After the taking of Prague on September 16, 1744, usable portions of the Austrian occupation forces entered Prussian service; 176 of them joined the regiment. In the process of doubling the Silesian garrison regiments as of July 8, 1754, twenty regiments in the central provinces were empowered on November 8, 1755 to recruit twenty men apiece from the Empire and "pay them ten Taler apiece". Thus eighty supernumeraries came to the regiment, plus twenty from the New Garrison Regiment, in all five junior officers and 500 men. The remainder of 845 men were to be found in Silesia. The funding began again as of June 1 of that year, the formation on August 15. The regiment supplied the wartime occupation of the Fortress of Glatz, a very important location. When the danger of war became more acute, the King extended the formation time to August 1, involving division of the existing companies, which were filled with recruits. The garrisons of both battalions were Glatz, Neisse, Habelschwerdt, Falkenhain (ten kilometers west of Glatz on the Reinerz-Habelschwerdt road) and Freiburg from 1743 to 1752, from 1753 to 1756 the same places minus Falkenhain and plus Bolkenhain, thus spreading them farther to the west along the mountains. Replacements came from the draft areas of I.R. 28 in Brieg, 33 in Glatz, 38 in Liegnitz and 49 in Neisse.

When a 4000-man corps of Austrian light troops advanced over the Landeshut Pass into Silesia after the defeat at Kolin in 1757, Major General von Kreytzen attacked them with six battalions from the forces occupying the fort, but was struck a hard blow on August 14 of that year. The majority of the regiment served in and around Glatz, securing that important county and its passes to Bohemia. Thus it was not used on field duty. The battalions' hour struck when the Austrians surrounded Glatz in July of 1760 with 13,000 men under Draskovich. "Since the regiment, consisting for the most part of prisoners and turncoats, did a bad job", the enemy was able to overrun the "Old Fortress" in the northern part of the city, on the cliffs high above the Neisse. Thereupon the commandant surrendered the city, the 3200-man occupying forces, and the "New Fortress" on the Schäferberg east of the Neisse. After Laudon's breakthrough at Landeshut, this was the second blow that led to the siege of breslau. During the Seven Years' War the regiment was not reconstituted; soldiers were lacking anyway, and prisoners were no longer exchanged.

Once the Austrians had evacuated Glatz on March 18, 1763, the regiment was set up anew under Colonel de le Noble, the former commander of a free battalion, with four battalions of the following battle-tested free troops: the Free Battalions of le Noble (since 1756) and Lüderitz (since 1758), and the Free Regiment of Wunsch (since 1758-1759), consisting of French and Austrian prisoners, all of whom had returned from imprisonment. From 1764 to 1788 three battalions were stationed in Glatz and one in Reichenbach. In 1788 the regiment was divided among the new Fusilier Battalion No. 7 and, with three companies each, I.R. 12 in Prenzlau, 24 in Frankfurt/Oder, 34 in Ruppin and 35 in Spandau, plus two companies to I.R. 37 in Glogau and 43 in Liegnitz to form depot battalions. They ceased to exist in 1806 as third battalions of these regiments.

Garnison-Regiment
Nr. VIII

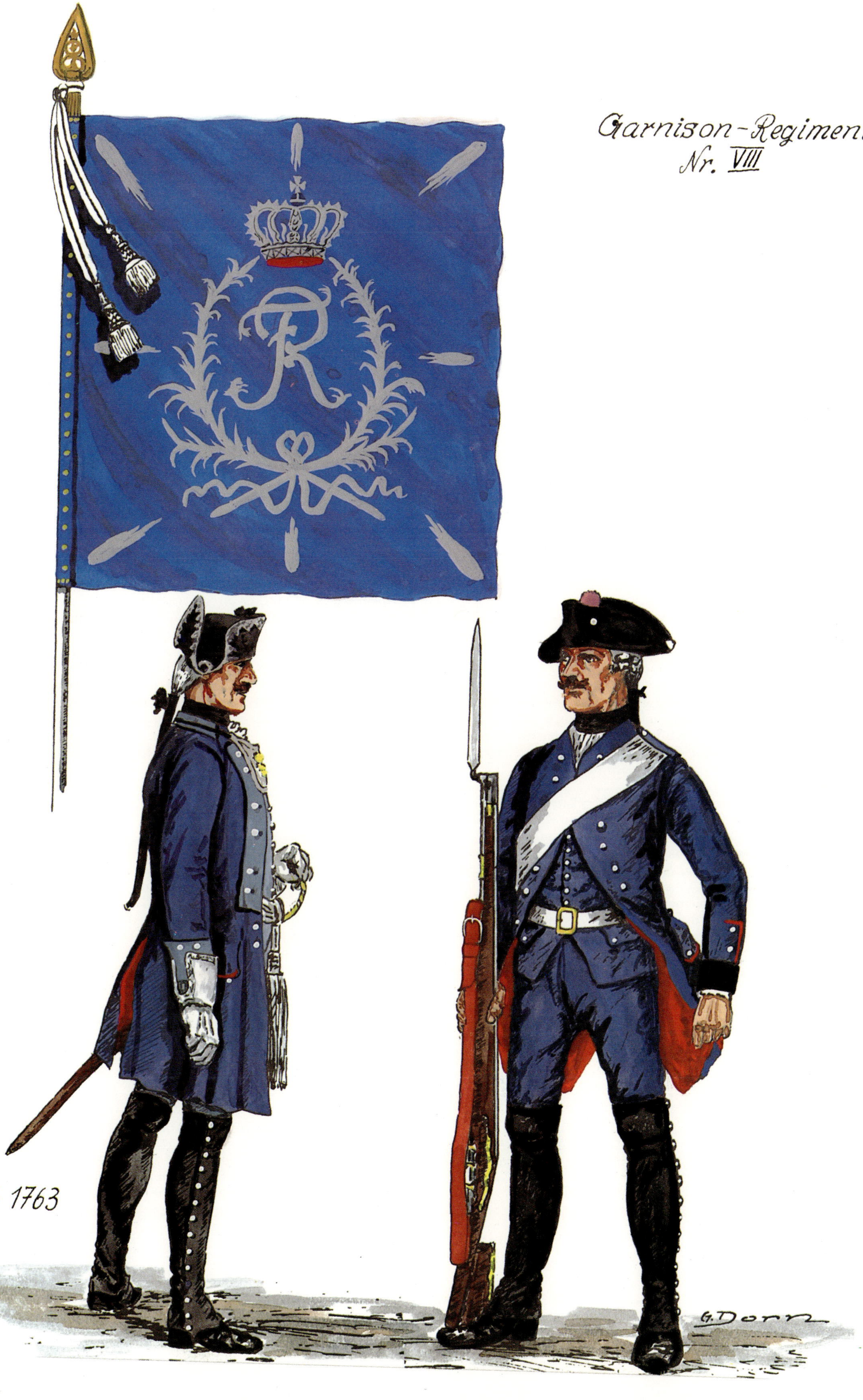

b 1763

GARRISON REGIMENT NO. IX

Commanders of the Regiment

1743	8/1	Major General Georg Volrath von Kröcher
1748	11/4	Lieutenant General Ernst August de la Chevallerie, Baron de la Motte
1759	1/17	Colonel Bernd Eckard von Bonin
1763	4/5	Major General Constantin Nathanael von Salenmon

The battalion was organized in Geldern on August 1, 1743, composed of the men of Kröcher's Battalion left over when most of them joined Fusilier Regiment 32 and marched to Silesia. Ten men of each company had been left behind as a core. The ranks were filled with recruits from the Geldern area. General von Kröcher was given command of the battalion. Geldern had become a western possession of Prussia in 1713, but since 1734-1735 and Friedrich II's attack on Silesia it had lost significance. This explains why the battalion was not enlarged until 1763. Prussia had turned to the east after Russia's appearance on the scene because of Austria and Poland. In Geldern one simply stayed in place; the decisive events happened elsewhere. There were no special rights for Geldern as there were for the Wesel garrison; until 1756 it was simply the garrison of the battalion. In the first half of 1756 it had to find and turn over sixty foreign recruits for the founding of I.R. 14. Until 1788 it had no draft area of its own, but recruited its own replacements from the Geldern area between the lower Rhine and the Maas. After Viersen was freed from the draft in 1736 by paying a fee, the King guaranteed four other towns freedom from enlistment for 800-Taler fees. After negotiations beginning on June 6 of that year, the senates of Geldern, Cleves and Tecklenburg agreed on May 24, 1748 to pay 30,200 Taler yearly for the removal of the draft areas. The battalion received 2600 Taler of this annually, which meant that it could recruit about 200 to 260 men per year.

The Seven Years' War disturbed the battalion when in 1757 France joined the alliance against Prussia. In April the French Army began to advance with more than 100,000 men. When Wesel was evacuated on March 24 of that year, according to plan, Geldern was isolated. The fortress, though surrounded by General Count Beausobre, was well protected by flooding, and Colonel von Salmuth was able to defend it successfully until August 23 of that year, longest of all forts in the west. Unreliable foreigners attempted a mutiny, which was put down energetically. When all food supplies were exhausted, Salmuth worked out the capitulation under terms of free departure. Once they left, the men scattered to the extent that on October 4 only fifteen officers, eighteen non-commissioned officers, twenty-four privates, one drummer and five flags arrived at Magdeburg. On October 13 they marched to Berlin, arriving in time to protect Potsdam from Hadik's attack. In the spring of 1758 the battalion returned to Magdeburg under its commander, Major von Rhaden, to be set up there. After peace was made it was sent back to Geldern and united with de Salenmon's Free Regiment to make two battalions. Major General von Salenmon was its chief for 25 years. Unlike the other garrison troops, it retained the sabers and light blue cuffs, collars and underclothes of the field regiments. In 1788 it provided the depot battalions of I.R. 44 and 48 in Wesel, which surrendered as battalions in Hamelin in 1806.

GARRISON REGIMENT NO. X

Commanders of the Regiment

1742	7/20	Colonel Georg Ewald von Puttkamer
1744	2/1	Major General Arnold Goswin von Rettberg
1747	9/5	Colonel Busso Christian von Blanckensee
1765	11/21	Colonel Christoph Ludwig von der Mülbe
1780	5/28	Colonel Heinrich Gottlieb von Könitz, later of I.R. 17
1786	3/1	Major General Carl Albrecht Friedrich von Raumer, formerly of I.R. 53, later of I.R. 36
1786	3/15	Colonel Friedrich von Oven

After the taking of Glatz on January 9, 1742 the King commanded on February 5 that recruits be sought in Bohemia for the formation of a garrison regiment. Colonel Georg Ewald von Puttkamer, who had been recruiting since October of 1741, was empowered to take charge as of July 20. Starting in Müncheberg on December 1, 1743 he began to build around small teams of the regiment taken over from Holstein-Gottorp, later I.R. 47 in Burg. On February 1, 1744 Major General von Rettberg became its commander. The garrison regiments lacked chaplains, gunsmiths and armorers, regimental drummers and oboists, as well as the fourth drummer in the bodyguard company of the second battalion and the two musketeer fifers. From 1744 to 1747 both battalions were stationed in Glatz, then until 1756 in Patschkau, Ottmachau and Nimptsch, and at times in Münsterberg as well. Replacements came from the Glatz area until 1747, then from I.R. 32 in Neisse, 38 in Liegnitz, 40 in Breslau and 42 in Reichenbach. In view of the threat of war, the King commanded on June 19, 1756 that the regiment be strengthened by two battalions without grenadiers, which was quickly done as of August 1 of that year. The Third and Fourth Battalions were intended for Neisse.

The "Department of Garrison Regiments and Battalions" was subordinate to the King's first Adjutant General. Major General von Tresckow was the Commandant of the Fortress of Neisse. In 1756 some parts of the fortress were still being finished, such as the palisades which would avoid the danger of storming. Neisse was an important magazine; here, for example, were stored 4.95 million pounds of flour in 450-pound barrels, which could be kept for forty years. Nearby a large siege park had been prepared, such as had been added at Olmütz but had been used against Schweidnitz. In 1758 it required four marching units for its 85 guns. Unlike Schweidnitz and Glatz, Neisse was not captured, even though it was heavily besieged from October 3 to November 7, 1758, after Schweidnitz had fallen on November 12, 1757 and the King's unfortunate advance on Olmütz had set out from Neisse in the spring of 1758. In 1762 the remaining men of I.R. 9 and 39, despoiled in Pomerania, came to the fortress, which was protected by Werner's Corps in the field. Dominance on the field and possession of the forts complemented each other. In 1763 the regiment received its four battalions. From 1764 to 1788 it was stationed in Neisse and Ottmachau, until 1777 also in Patschkau and Nimptsch. In 1788 it provided Fusilier Battalion 13 and three companies apiece as depot battalions for I.R. 33 in Glatz, 40 in Schweidnitz, 42 in Breslau, 47 in Crossen and Züllichau and 49 in Neisse. One company joined I.R. 37 in Glogau. In 1806 Fusilier Battalion capitulated, as did the third battalions of I.R. 40, 49 and 37. The musketeer battalions of I.R. 33 and 47 outlasted the catastrophe in Silberberg and joined the Grenadier Regiment King Friedrich III (Second Silesian) No. 11 in Breslau. The Third Battalion of I.R. 42 stayed in East Prussia and joined Grenadier Regiment King Friedrich Wilhelm II (First Silesian) No. 10 of Schweidnitz. The grenadiers formed the Fusilier Battalion No. 10, which had to surrender at Jena.
Uniform and flag illustrated on page 158.

Garnison-Regiment
Nr. IX
ab 1763
G. Dorn

Garnison-Regiment
Nr. X

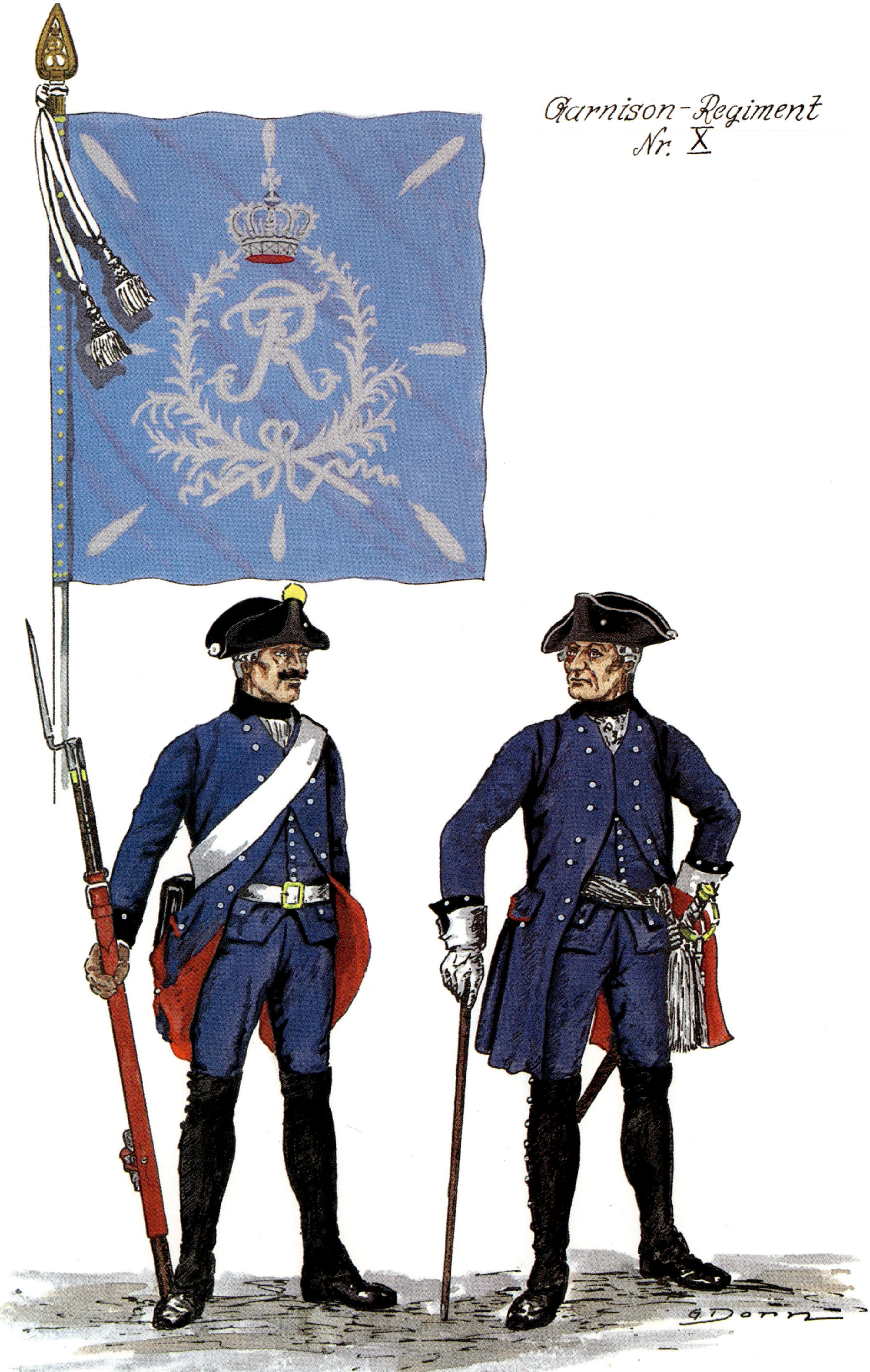

GARRISON REGIMENT NO. XI

Commanders of the Regiment

1744	1/25	Colonel Georg Ewald von Puttkamer
1748	9/6	Colonel Franz Christoph von Manteuffel
1760	2/5	Colonel Henning Christian von Mellin
1769	3/26	Colonel Carl Ludwig von Ingersleben
1782	1/5	Colonel Sigismund August von Berrenhauer

In addition to Garrison Regiments I and II, a third regiment of two battalions was formed between November 1743 and January 1744 of men from the two regiments plus recruits. Each regiments contributed a grenadier company. The staff went to Angerburg, bringing one battalion with them, while the other went to Königsberg/Pr. On January 25, 1744 Colonel Georg Ewald von Puttkamer, formerly of G.R. X, became its commander. After the taking of Prague on September 12 of that year, it was filled with 465 men from the occupation force of war prisoners. On July 27, 1756 Field Marshal von Lehwaldt received orders to enlarge the regiment to four battalions. On September 15 it was stationed in Heiligenbeil, Liebstadt and Preussisch Holland, without grenadiers. The regiment supplied 150 men, I.R. 2, 4 and 14 each 300 soldiers, and Dragoon Regiments 9 and 10 each 150 men, 1350 in all. From 1744 to 1753 it was stationed in Angerburg and Königsberg, from 1753 to 1756 in Heiligenbeil, Kreuzburg/Pr., Zinten and Domnau. Replacements came from Angerburg, Barten, Lötzen, Johannisburg and Lyck from 1744 to 1763. In 1756 the regiment kept its furloughed men in the garrison.

At Gross-Jägersdorf on August 30, 1757, it fought on the right wing of the second line. Early in December of that year the regiment marched to Silesia to take part in the surrounding of Schweidnitz. After the surrender on April 16, it stayed there as an occupation force. In 1760 all four battalions were on the field budget and joined Fouqué's Corps. Only the Fourth Battalion stayed with Fouqué; the Second and Third went to Freiburg in Zieten's Detachment. The First Battalion was sent to Neisse. In the night of July 23 the Fourth Battalion was taken prisoner at Landeshut. Zieten led the remaining men to Breslau via Schweidnitz. After Liegnitz on August 15, the Second and Third Battalions marched to Neisse, which had to be held. The first three battalions stayed there; the Fourth was not reformed. When peace was made, six companies stayed in Heiligenbeil until 1782, three each in Cruzburg, Zinten and Preussisch Eylau. Replacements came from the districts of Seehausen, Rhein, Ortelsburg and Drigallen, with Nikolaiken, Arys and Bialla. Because of West Prussia the Fourth Battalion was reestablished in Mühlhausen, beginning on October 1, 1772. From 1782 on one battalion each was stationed in Heiligenbeil and Heilsberg, three companies each in Guttstadt and Allenstein, two companies each in Wormditt and Wartenberg.

In 1788 the regiment formed Fusilier Battalion 11 in Heilsberg, the depot battalion of I.R. 7 in Stettin, the Third Battalions of I.R. 51 in Marienburg, I.R. 53 in Braunsberg, I.R. 54 in Graudenz and I.R. 55 in Preussisch Stargard, which fought in Danzig and Graudenz in 1806, later joining Grenadier Regiment Count Kleist von Nollendorf (First West Prussian) No. 6 of Görlitz and Grenadier Regiment King Wilhelm I (Second West Prussian) No. 7 of Liegnitz.
Uniform and flag illustrated on page 160.

GARRISON REGIMENT NO. XII

Commanders of the Regiment

1744	9/6	Colonel Ernst Georg von Kalckreuth
1778	11/5	Colonel Guillaume René de L'Homme de Courbière, later commander of a free regiment
1787	6/14	Lieutenant Colonel Philipp Lambert de Bauvrye

Despite the inheritance agreement concluded in 1691 between the East Frisian princely house of Cirksena and the Guelphs, Emperor Leopold I gave Brandenburg the Imperial reversion of East Friesland. Troops from Brandenburg had been in Emden and Greetsiel since 1682, two companies of "Mariniers" which were part of the "New Garrison Regiment" from 1742 to 1744. When Prince Carl Edzard died, young and childless, on May 25, 1744, the King ordered a detachment from I.R. 28 in Wesel, commanded by Colonel the Count of Wied, to march in. The Prince's widow, born a countess of Bayreuth, was related to the King. Next the Prince's four guard companies stationed in Aurich were taken into Prussian service, converted to three "on the Prussian foot" and united with the two Prussian companies in Emden and Greetsiel to make one battalion under Colonel von Kalckreuth, but without grenadiers. The two Prussian companies had eight officers, twenty non-commissioned officers, six drummers, two medics and 224 privates. The three new companies had nine officers, 24 non-commissioned officers, nine drummers, three medics and 240 privates. This added up to a battalion of 17 officers, 44 junior officers, 15 drummers, five medics and 464 men, thus a weak one. It was now called the "East Frisian Garrison" or "Garrison Company in East Friesland". On September 6, 1744 Kalckreuth was given command of it. The battalion was stationed at an outpost, 150 to 250 kilometers away fromPrussian territory as the crow flies, and now meaningless after The Netherlands were no longer a great power and Prussia lacked any sea power. The companies were separated in their posts until 1750, then from 1751 to 1788 they were together in Emden. It had no draft area, but drew its replacements from the Emden-Aurich area. Even though the Prussian Army was not very familiar in East Friesland, the small garrison was respected, particularly now that the hostility between the parliament, supported by The Netherlands, and the Prince had ceased. Since Kalckreuth stayed there as commander for 34 years, the battalion was satirically called "die Kalckreutter" (the chalk riders) in the local dialect.

When the French began their advance in April of 1757, took possession of Wesel and surrounded Geldern, Colonel von Kalckreuth surrendered Emden to the detachment of General the Marquis d'Auvet, which was to occupy East Friesland. The battalion had proved itself to be "completely unreliable", since it consisted of 'foreigners'. This judgment may be questioned—perhaps it was simply much too weak. The battalion was exchanged in 1759, with most of the officers going to I.R. 45 of the "Wesel Garrison" or the "Volontaures de Prusse" free battalion in Minden, led by Major von Trümbach. It proved itself under Duke Ferdinand of Braunschweig in the "Allied Army". It also contributed men to Garrison Regiment IX in Magdeburg and to the artillery. In 1763 the First Battalion of the Courbire took its place in Emden. Colonel Guillaume René de L'Homme de Courbière, a Huguenot from The Netherlands, now became its chief, until then commander. He became famous as the defender of Graudenz in 1806. In 1788 it was disbanded to form the depot battalions of I.R. 10 in Bielefeld, 45 and 48 in Wesel, which capitulated in Nienburg, Erfurt and Hamelin in 1806.
Uniform and flag illustrated on page 3.

Garnison-Regiment
Nr. XI